SOLID WASTE DISPOSAL

Volume 1

Incineration and Landfill

SOLID WASTE DISPOSAL

Volume 1

Incineration and Landfill

BERNARD BAUM, PhD, Vice President and Manager
of Materials Research & Development Division
    and
CHARLES H. PARKER, Consulting Associate

DeBELL & RICHARDSON, Inc., Contract Research &
Development, Enfield, Connecticut

The original work for these volumes was
based on work carried out under a contract with the Manufacturing Chemists
Association.

ann arbor science PUBLISHERS INC.

POST OFFICE BOX 1425 ● ANN ARBOR, MICHIGAN 48106

Copyright © 1973 by Ann Arbor Science Publishers, Inc.
P.O. Box 1425, Ann Arbor, Michigan 48106

Library of Congress Catalog Card No. 73-82272
ISBN 0-250-40034-0

Printed in the United States of America

PREFACE

In 1970, national policy took three steps toward
involvement in the management of the environment.
These three steps were: (1) creation of the Federal
Environmental Protection Agency (EPA); (2) establishment of a Council on Environmental Quality (CEQ) in
the Executive Office of the President; and (3) passage
of the National Environmental Policy Act. EPA has
major responsibilities in administering and enforcing
Federal antipollution laws in air, water, solid
wastes, noise, pesticides and radiation areas. CEQ
has responsibilities in developing national environmental policy and new programs and in coordinating
the federal environmental effort. CEQ also prepares
an annual report on environmental quality. The
National Environmental Policy Act's primary purpose
is to ensure that the environmental aspects of every
major federal-agency decision are evaluated fully
and promulgated publicly.

Even with the establishment of the above aids
to national environmental policy, cooperation from
state and local governments in adopting similar
legislative, administrative and funding efforts will
be required. Further, the private sector must also
take initiative in developing and complying with
reasonable pollution abatement standards. Without
such an integrated effort for managing environmental
quality, it is highly questionable that any plan or
law or set of laws, even backed up by enormous sums
of money, can attain any level of success.

It would be an utopian condition if management
of environmental quality, in all of its facets of
government, business and the public, did not produce
inevitable conflicts of viewpoints. In order to
minimize such conflicting views, governmental influence should provide intergovernmental *modus
operandi* on the three levels of federal, state and
local jurisdiction to assure effective coordination
of policies involving planning and financing.

Further, possibly a joint public-private effort
to combine the technology and other resources of
corporations and local governments to combat
environmental problems should be provided.

In 1971 it is estimated that, in the United
States alone, there were discarded: 70 billion
metal cans; 38 billion glass containers; more than
7 million television sets; nearly 3.7 million tons
of plastics; more than 100 million tons of paper
products; and several million junk automobiles.
The manner in which we live has contributed to these
mountains of waste and to the problems of disposal.
Solid waste is generated by society in the following
approximate proportions: 44% from private families
or households; 30% from construction activities and
industry; and the remaining 26% from commercial
establishments, such as stores and warehouses.
Packaging contributes about 25% of the approximately
200 million tons of solid waste reported as collected
in 1970. The food industry is considered the largest
user of packages, followed by beverage and chemical
product companies. Packaging materials in 1966 were
estimated at about 52 million tons and are expected
to rise to about 74 million tons of waste in 1976.
Difficulties of disposing of such materials may
become more complex because dissimilar materials
are often combined in the same package.

The United States spent, in 1970, $5.7 billion
on solid waste disposal. CEQ estimates that the
country will have to spend $7.8 billion in 1975--
over 35% more than in 1970. For the years between
1970 and 1975, CEQ places the total cost of solid
waste disposal for the period at $43.5 billion. In
choosing a method for solid waste disposal for the
community or other unit, some factors should be
taken into consideration: (1) make the best possible
use of natural resources, either recovering energy
or product; (2) avoid the creation of some other
form of pollution which may generate more hazard
than is present in the original waste; (3) obtain
the best balance of economy and efficiency. New
methods of waste management vary widely because no
two communities have the same problems--so no single
technique will work in every case.

Two methods account for about 15% of the solid
waste disposed of in the United States--almost all
of the balance is disposed of simply by open dumping.
The two methods discussed in detail herein are:
(1) *sanitary landfill*, and (2) *incineration*. A
sanitary landfill is not a dump--it is a smokeless,

odorless, ratless engineering project. An incinerator, designed to combust municipal refuse, is also
an engineering project. Properly designed, constructed, administered and operated incinerators are very
effective in reducing the total volume of municipal
refuse by up to 90%. Other disposal methods presently
are not used widely enough to be compared with
sanitary landfill and incineration as able to handle
the quantities of refuse generated in urban areas.

Bernard Baum
Charles H. Parker
June 1973

CONTENTS

PART I

STATISTICS AND COLLECTION

PART II

INCINERATION

PART III

SANITARY LANDFILL

PART IV

PLASTICS IN INCINERATION AND LANDFILL PROCESSES

LIST OF TABLES

Table

Table

LIST OF FIGURES

Figure

Figure

PART I

STATISTICS AND COLLECTION

CHAPTER 1

SOLID WASTE STATISTICS

NATIONAL SURVEYS

National surveys[1,2] show that the average amount
of solid waste actually collected in the United
States in 1968, the most recent year for availability
of such data, was about 5.3 pounds per person per
day, 1935 pounds per person per year, or more than
190 million tons per year. Waste generation could
logically be expected to follow the rise in per
capita purchases for nondurable and durable goods--
during the past few years this represents an annual
increase of about 4%. Thus, the amount of collected
refuse, through private as well as municipal collection agencies, should rise to 8 pounds per capita
per day by 1980, or 2920 pounds per person per year.
Considering the increase in national population by
1980, this means over 340 million tons of solid
waste can be expected to be collected in that year.
These figures represent only the amount of
material which is handled by public and private
collectors. Ten to fifteen percent of household
and commercial wastes are self-collected and transported to disposal sites. About 30-40% of the
industrial wastes are also self-collected and
transported. When these estimates are added to the
known figures for collected refuse, the present
total amount is nearly double the 190 million tons
calculated for 1968. Estimates for 1968 indicate
that over 10 pounds per day of household, commercial
and industrial wastes were generated for every man,
woman and child in the U.S., totaling over 360 million
tons for the year. The 1968 figures can be broken
down to 250 million tons per year of household, commercial and other nonindustrial wastes to which must
be added the national survey estimate of 110 million

3

tons per year for industrial solid wastes. Figures
for 1970-1971 are estimated at 390 million tons.

To obtain the total of over 3.5 billion tons of
solid wastes generated in the U.S. in 1968, the
figures show a breakdown of 550 million tons of
agricultural and crop residues, 1.5 billion tons of
animal wastes, 1.1 billion tons of mineral waste,
and the previously mentioned figure of 360 million
tons of generated household, commercial and
industrial solid wastes.

Further figures indicate that of the 5.3 pounds
of solid wastes *collected* for each person in the
U.S. in 1968, about 3 pounds is known to be household in origin, 1 pound commercial, 0.59 pound
industrial, 0.18 pound demolition and construction
and 0.55 pound miscellaneous. These figures include
only material known to be collected. Household,
commercial, industrial, demolition and other solid
wastes that were transported to disposal sites or
disposed of by the generating party are not included.

According to the American Public Works Association, the U.S. population has increased 30% since
1950, but the waste load has increased 60% and is
expected to rise another 50% by 1980.

Projections for Collected Refuse
Including Plastics

The following data show the growth in "collectable
refuse" (residential, commercial and industrial
wastes--excluding agricultural and mineral wastes)
in the United States, from a consensus of various
sources:

	1950	*1960*	*1970*	*1980*
Total (million tons)	105	135	195	230

Figure 1.1 shows the percentage collected by
public, private and individual sectors of household,
commercial and industrial solid wastes for 1968. By
1980 the plastics share of this solid waste figure
is expected to be about 2.8% or 6.44 million tons,
up from 2% or 3.9 million tons in 1970 in the United
States. Estimates[3] indicate about 55% of this is
polyolefin, about 20% polystyrene, about 11% PVC
and 14% all other, including cellulosics. For
packaging wastes only, polyolefins account for
nearly 75% and polystyrene nearly 20%. In industrial

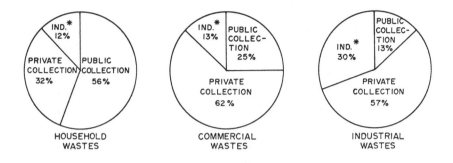

(POPULATION BASIS)

*Figure 1.1.   Performance of solid waste collection in the U.S.
(1968).*[2] *Collection by individual householders,
institutions, or industries.*

wastes, plastics contributed about 0.5 million tons
in 1968, or about 12% of the plastic waste totals--
0.5% of the 110 million tons total industrial wastes.

Road-Side Litter Survey

A recent survey by the Highway Research Board of
the National Research Council[4] listed the following
litter along a one-mile stretch of ordinary two-lane
highway: 770 paper cups, 730 empty cigarette packs,
590 beer cans, 130 soft drink bottles, 120 beer
bottles, 110 whiskey bottles and 90 beer cartons.
Other choice items in this collection include bed
springs and shoes. This is termed *solid waste* by
the layman, but it is actually *litter* to those
persons who recognize that litter is but one
component of our solid waste disposal problem.
In a study of roadside litter, the following
data are shown:

Paper Item Sub-Class	Items per Mile	% of Total
Newspapers or Magazines	25	1.89
Paper Packages or Containers	150	11.52
Other Paper Items	601	46.08
Total	776	59.49

Item Class	Items per Mile	% of Total
Paper Items	776	59.49
Cans	213	16.31
Plastic Items	75	5.78
Miscellaneous Items	163	12.53
Bottles and Jars	77	5.88
Total	1,304	99.99

It was estimated that one cubic yard of litter was accumulated per month, on the average, for each mile of interstate and primary highway in the 29 participating states for the calendar period represented by the pickups. An estimated 59% of all items were paper, 16% were cans, 6% were plastics, 13% were miscellaneous and 6% were glass jars. It appeared that total litter volume was positively correlated with average daily traffic, but no other relationships were clearly established with such roadway factors as right-of-way width or type of roadside cover.

Total Disposal Figures

The total residential, commercial, and industrial waste collected by municipal and private contractor facilities in 1968 was: 5.3 lb/person/day x 200,000,000 people = 1.060 billion pounds per day.

At present, landfill is by far the most common method of solid waste disposal in cities having populations of 25,000 and over. Figure 1.2 shows this dramatically. There are at least 75,000 tons per day of incineration capacity installed at the present time. If an average utilization of 60% and an average solid waste generation figure of 5.3 pounds per capita per day are assumed, the average percentage of waste incinerated amounts to 9%. The percentage of waste disposed of in composting plants and by other methods is negligible.

The annual 1971 total of solid waste and litter includes 71 billion cans, 38 billion bottles and jars, 7.6 million television sets and 7 million automobiles and trucks. The total also includes about 3.8 million tons of plastics and 40 million tons of paper. By 1976, the dry weight of plastic solid wastes is predicted at a little over 5 million tons. Based on the weight of refuse collected, the percentage of plastics is between 1.5% and 1.7%, on

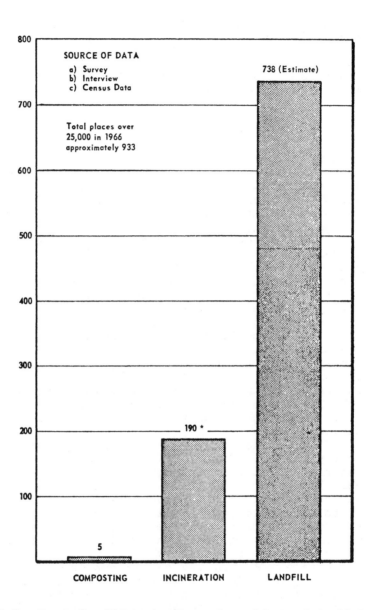

Figure 1.2. *Type of solid waste disposal used by municipalities of over 25,000 population--1966.*[5] *Some cities have more than one incinerator installed. There are a total of 250 incinerator plants in the U.S. Some incinerators handle refuse from several cities.*

the average, and may be expected to rise to between
2.2% and 2.4% by 1976. It is not expected to exceed
3% by 1980. At the present time, all nonchlorine-
(or other halogen-) containing plastics comprise
about 88 to 90% of the total plastics which enter
the solid waste stream.

Plastic Solid Waste in Packaging

     According to a survey made for the U.S. Depart-
ment of Health, Education and Welfare by Midwest
Research Institute on "The Role of Packaging in
Solid Waste Management--1966 to 1975," the packaging
industry contributed 51.7 million tons to the total
of residential, commercial and industrial waste in
the U.S. in 1966. Further statistics indicate that
only about 10% of this total failed to reach disposal
sites during the year produced. The remaining 90%
would then constitute nearly 25% of the 0.2 billion-
ton figure for all "collectable refuse" mentioned
earlier. This survey also indicates that about 1.1
million tons of plastics were included in this total
of 51.7 million tons, or approximately 2%. The
plastics total includes film, sheet, bottles, tubes,
and all other formed containers, trays, crates and
closures. It also includes all types of resins used
in the packaging industry (including cellophane and
other cellulosics) but does not include an estimated
0.237 million tons of resins used as coatings for
packaging.
     It is conservatively estimated by this same
survey that the contribution from all packaging
materials will rise from the 51.7 million tons
calculated for 1966 to 73.5 million tons in 1976,
an increase of 21.8 million tons. About one-third
of this increase, 6.9 million tons, will be accounted
for by population increase; about two-thirds of the
increase, 14.9 million tons, will be generated by
changing customer habits which increase consumption
per capita. Actual per capita consumption of
packaging materials in 1966 was 575 pounds. This
is expected to increase to 661 pounds by 1976. In
addition to this increase, it is expected that consumption of plastics packaging materials will
increase enough over that of other materials to
reach about 8% of the total, or about 5.8 million
tons.

Disposal Costs

The management of solid wastes, including plastics, involves several steps--collection, sorting, storage, processing and disposal (which includes transportation). The average cost of refuse management in the U.S. is expected to rise from the 1969 level of about $3.7 billion ($21 per ton) to about $6 billion ($27 per ton) in 1972. In a survey of 137 of its member companies, Manufacturing Chemists Association (MCA) found that, in 1972, these reporting members expended $83 million in capital investment (original installed cost) for environmental solid waste management facilities which are continuing to be in use in the production of chemical materials, some of which are plastics. The survey also showed projected added capital investment, 1972 through 1975, of $92 million in the field of solid waste management. Other figures shown in the survey for solid waste management were $37 million for annual operation and maintenance costs of the facilities and $3.4 million annually for research expenditures.

Less specific to any particular industry, private group or municipality, an informative report released August 7, 1972, is the Council of Environmental Quality's (CEQ) Third Annual Report to the President and Congress. For the decade 1971-1980, cumulative cash expenditures for solid waste management will total $86.1 billion. These ten-year expenditures are made up of an estimated $3.0 billion in capital investment and $83.1 billion for operating costs. This $86.1 billion in cash flow is about 30% of the total 10-year projected expenditure of $287.1 billion for all pollution control.

Aside from costs, this particular report projects a municipal solid waste generation of 810 million tons per year by the year 2000, a 225% increase over current levels, with a management cost to all local governments of $15 billion per year.

REFERENCES

1.   Office of Solid Waste Management, Environmental Health Service, Department of Health, Education and Welfare. *Massive Statistics*. Public Health Service Publication No. 1867 (1968).
2.   Office of Solid Waste Management, Environmental Health Service, Department of Health, Education and Welfare. *Interim Report*. *1968 National Survey of Community Solid Waste Practices*.

3. Warner, A. J., C. H. Parker, and B. Baum. "Solid Waste
    Management of Plastics," for Manufacturing Chemists
    Association, p. A-11 (December, 1970).
4. *National Study on Roadside Litter*, National Research
    Council, Highway Research Board, pp. 8-12.
5. Office of Solid Waste Management Programs (OSWMP),
    Environmental Protection Agency (EPA). "Technical
    Economic Study of Solid Waste Disposal Needs and Prac-
    tices," compiled by Combustion Engineering, Inc.
    Public Health Service Publication No. 1886, *Vol. I,
    Municipal Inventory* (W. R. Copp).

Unreferenced material found in this two-volume series
is information from the voluminous files of authors
Baum and Parker and the records developed by DeBell
and Richardson, Inc. in the consulting organization's
studies of solid waste.

CHAPTER 2

COLLECTION AND SORTING

COLLECTION

The collection of solid wastes involves storage at the place of origin and transportation to the point of disposal. The method of collection of refuse is related to the method of disposal.

The collection of municipal refuse is estimated to constitute 75% of the disposal costs according to some municipal authorities. Considerable development work on collection methods is presently under way. Compaction at the point of collection or enroute to a major storage point has assisted in reducing volume prior to transfer to the disposal site. Vehicles are largely (69%) compactor trucks, 8% enclosed noncompactor trucks, and 23% open dump trucks. Costs of the vehicles ranged from $13,000 to $30,000, depending on size, type and auxiliary equipment furnished. For about 15 million people in 47 communities, 30,000 to 40,000 vehicles would be needed exclusively for waste collections.[1]

The hauling of refuse to the disposal site is done in many kinds of vehicles. These include open trucks as well as specially designed equipment which has an enclosed liquid-tight storage body and provides for some compaction of the collected materials.

The final disposal of most of the solid wastes is still accomplished in four basic ways: open dump, sanitary landfill, incineration, and by salvage, which includes composting and hog feeding. In addition, some of the garbage ground at the points of origin is disposed of through the sewage system.

A survey of 995 communities with 5,000 or more inhabitants shows the following structure of refuse collection practices according to type of collection organization.[1]

Collection Organization	*Percent Share of Total Number of Communities*
Municipal	44.3%
Contract	17.6
Private	13.1
Municipal and Contract	3.3
Municipal and Private	15.2
Municipal, Contract and Private	1.6
Contract and Private	4.4
Unknown	0.5

Control of Collection

The data indicate that in 65.2% of the communities the control of solid waste collection was vested completely in public authorities through either municipally-owned or contractually-arranged operations. It must be kept in mind that these data do not cover the practices of smaller communities and unincorporated areas for which published data are not available. It is estimated that in the majority of small communities with populations of less than 5,000, the disposal is handled by private companies or by the producers of the wastes.

Collection Costs

The current cost for suitable refuse collection equipment is estimated as follows:[1]

Regular compactor trucks	10-20 cubic yard $10,000 to $13,000 each
Heavy-duty compactor trucks	24-28 cubic yard $15,000 to $20,000 each
Trailers	30-80 cubic yard $25,000 to $30,000 each
Compactor trucks with detachable container and hoisting unit	$15,000 to $30,000 each

Collection accounts for the bulk of refuse removal costs. Ranging from $5.00 to $25.00 per ton, collection costs are commonly estimated to make up 65-80% of total disposal cost. Transportation cost, excluding depreciation of equipment, of a typical 18-22 cubic yard packer truck carrying from three to four tons of compacted refuse, is estimated at $0.35 to $0.40 per mile. The average trip is estimated to be between 10 and 25 miles.

In consideration of charges for these operations, according to APWA surveys, it is estimated that 35-36% of the communities finance their refuse collection and disposal operations through service charges, 50-52% through general taxes, and 12-15% through a combination of taxes and service charges.

The number of collection vehicles in use is extremely hard to estimate. All kinds of vehicles are used and many of the vehicles serve other purposes as well. Of the government-owned collection vehicles, 68.5% are estimated to be compactor trucks, 8.3% enclosed noncompact trucks, and 23.2% open dump trucks. The capital investment needs for collection and transfer equipment plus maintenance and storage facilities are estimated to be $1.42 billion during the 1966-1976 decade. This includes the replacement of almost all noncompactor types of existing truck fleets with compactor trucks costing an average of $13,000 each. About 34% of the total estimated capital investment needs for refuse collection facilities are needed for providing service where none currently exist or up-dating the current service to acceptable levels.

The collection and removal of municipal refuse-- one of the major problems of American cities--has been given less attention than this public function deserves. Only within recent years have most municipal officials been willing to admit that refuse is a technical management problem worthy of their attention and study. Much progress has been made, but still only a minority of communities are using the administrative techniques which generally have proved most satisfactory. Not that methods, equipment, and practices should be uniform over the country. Conditions vary and it is vital that procedures vary to meet them. However, the problem should be approached in the same way in all communities, should be analyzed in terms of sound administrative management, and should receive the same consideration usually given other public health aspects of government.[2]

Critical Needs for Collection

What are the critical areas of need for municipal collection systems? First, recommended criteria must

be developed for collection systems. Such criteria
can be developed to facilitate adoption of standards
by state and local governments. These criteria can
best be developed by the EPA's Office of Solid Waste
Management Programs with the assistance and cooperation of the state agencies and concerned associations,
such as the American Public Works Association. Manpower development, training and education also are
critical needs. Education is required for management,
equipment operation and collection personnel--and the
general public.

Solid wastes must be managed. This means development of systems which incorporate processing and
storage at the point of generation, transport to the
point or points of disposal, and the disposal process
itself. Each of these operations represents a marriage
of technique, equipment and people.[2]

SEGREGATION AND SORTING

A lot of money is being spent to find solutions
to the problem of how to separate usable materials
from raw refuse. Mechanized reclamation from municipal
solid waste currently is being investigated by researchers in the Mechanical Engineering Department
of Massachusetts Institute of Technology (MIT) in
Cambridge, Massachusetts. They have concluded that
the prospects for successful and economic reclamation
of the major components of domestic refuse are bright.

At present, the MIT research effort primarily is
involved with the separation of trash at central
treating plants. The researchers feel that reclamation
plants could be profitable now in some high disposal
cost areas. They have evaluated a variety of identification and sorting systems using techniques such
as magnetic coding and sorting, conductivity coding,
photometric and radiometric coding, air switching,
x-ray attenuation coding, vortex classifying and
many others. However, they feel a series system
with manual coding and mechanical switching offers
the greatest possibility of profit at present.
Such a system also permits evolutionary development.[3]

Aside from an increased reclamation mindedness,
one basic need may be the tagging of materials during
manufacture to permit better reclamation at the other
end of the use cycle when the product is discarded.
Such tagging also might allow tracing of polluting
materials in the environment. It would permit
easier identification of industrial and municipal

refuse. Governmental agencies would welcome this type of assistance in finding, investigating and prosecuting polluters.

A machine (shown in Figure 2.1)[4] which sorts by density and drag and which was developed for the food industry principally for the sorting of food beans, has been applied by Boettcher of Stanford Research Institute (SRI) to the separation of mixed municipal refuse.[5] The so-called air classifier

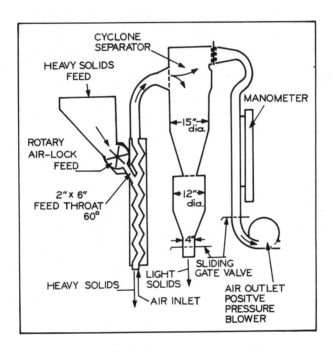

Figure 2.1. Flow diagram of air classifier as developed by Stanford Research Institute.[4]

consists of a zig-zag duct arranged with its principal axis vertical. Air is drawn off from the top and the mixed material is fed into the duct at an intermediate point. Pieces with a combination of low density and high drag are entrained into the rising air stream, while the heavier or lower-drag pieces descend. It is claimed by SRI that components having densities only a few percent apart can be

successfully and economically sorted. If this claim
is substantiated, the machine could obviously have
a very beneficial impact on the reclamation picture.
This type of machine requires the input to be pulverized, dried and screened so that any one classifier
is fed with pieces varying in size by only a small
degree. Figure 2.1 shows the flow scheme.

Cumberland Engineering Company's Waste Doctors
Division has described a method of separating plastics
and claims the ability to divide polymers with specific gravities as close as ± 2 lb/cu ft. The method
is called a mechanical vacuum-gravity separation; it
is claimed that this will easily separate polyvinyl
chloride (PVC) from polypropylene (PP) and mixtures
of three or more polymers provided that their specific
gravities differ by as much as ± 3%. Therefore, a
mixture of rigid PVC of specific gravity 1.4, polystyrene (PS) of specific gravity 1.06, and high
density polyethylene (HDPE) of specific gravity 0.95
could be divided easily; however, modifiers such as
pigments, fillers, extenders and stabilizers can
alter the specific gravity of materials and could
result in some confusion in the separation process
regardless of technique.

Jensen, Holman and Stephenson[6] describe some of
the work on plastics separation being done at the
Bureau of Mines. Very briefly, this involves sink-
float separation by density. The method separates
five types of plastics into clean fractions: starting
materials, high density polyethylene (HDPE), low
density polyethylene (LDPE), polypropylene (PP),
polystyrene (PS) and polyvinyl chloride (PVC).

Water divides the mixture by density as follows:

Plastic	Density
PVC	1.313 gm/cc
PS	1.055 gm/cc
HDPE	0.958 gm/cc
LDPE	0.916 gm/cc
PP	0.901 gm/cc

The PS and PVC are then divided in an aqueous
calcium chloride solution with a density of 1.28
gm/cc which floats the PS away from the PVC.
Alcohol-water solutions are used to further separate
the fraction that floats on water. A 50-50 solution,
with a density of 0.93, floats the LDPE and PP away
from the HDPE, and these two components, LDPE and PP,
will then separate in a solution of 60 parts alcohol

in 40 parts water. The application of this sink-
float method to municipal waste plastics appears to
have much potential, but additional study will be
required to determine the extent to which dirt,
attached labels and fillers in the waste plastics
may interfere with the separations.

The PE and PP fraction separated by a method
such as that outlined above, which would probably
constitute about three-fourths of the total plastics
in urban collections for the next few years, can be
fabricated into useful products by compression
molding or injection molding, but probably would not
be amenable to methods requiring close control of
flow properties as in blow molding.[6]

Figure 2.2 illustrates some of this work done
by the Bureau of Mines. The Bureau of Mines is
developing other methods for reclaiming the
valuable materials contained in unburned urban
refuse. One phase of this research deals with the
recovery of plastics from shredded refuse using an
electrodynamic separation process. The results of
this study have indicated that mixed plastics con-
centrates analyzed at 97% plastics can be recovered
from shredded refuse leaving a mixed paper fraction
analyzed at 99.8% paper. Recovery of the plastics
was over 97%.[8]

Kenahan and Sullivan[7] describe a third Bureau
of Mines process, this one for a method of sampling,
separation and classification with results obtained
from residues of several incinerators in metropolitan
Washington, D.C. (1967). The actual separation
procedure is described.

After drying of the wet residues, dry bulk
density measurements were made and separation of
the various fractions was begun. Large pieces of
materials including tin cans, massive iron, glass,
iron wire, paper, stones, bricks, ceramics, unburned
organics and nonferrous metals were hand-picked
directly from the trays. The remaining dry residues
were then processed through a two-deck oscillating
screen fitted with 2- and 8-mesh screens. Fractions
obtained from the screening process were: (1) coarse
material which was greater than 2-mesh, (2) inter-
mediate-size material (minus 2-, plus 8-mesh), and
(3) fines (minus 8-mesh). Each of these fractions
was passed over a permanent-type magnetic pulley
several times to remove the ferrous metals and
magnetic iron oxides. The larger pieces of nonferrous metals were hand-picked during the magnetic
separation. Both the course and intermediate

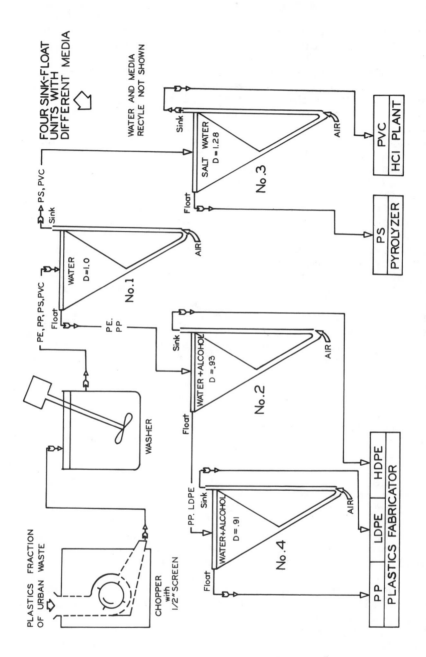

Figure 2.2.   Flow diagram of proposed experimental sink-float process for separating waste plastics.[6]

fractions were chiefly glass. These were processed in an air separator to separate the glass from lighter materials such as paper and ash. The intermediate fraction was then quartered in a sample splitter and one quarter was pulverized in a gyratory crusher. The larger pieces of non-ferrous metals were then hand-picked from the crushed portion. The crushed glass was then quartered to about a 5-pound sample, ball-milled and screened to separate the flattened pieces of nonferrous metals from the pulverized glass. During all processing operations, air-suspended dust and fines were recovered in a dust collector. All dust was combined with the remaining fine fraction, and the total was reported as ash.

Partly mechanized reclamation plants recovering up to 50% of the input of mixed municipal refuse can now be designed to be economically profitable in high-disposal-cost regions of the country. Many approaches currently being used in different institutions around the country promise to give greater economies in the future and thereby to allow the more widespread use of reclamation as a solid waste treatment method. Incentives are needed to increase the market for secondary materials and to encourage manufacturers to design reclamation into their materials, packaging, and products.[3-5,9,10]

REFERENCES

1. *American Public Works Reporter (APWR).* 5 (August, 1966).
2. Weaver, L. "News from the Institute for Solid Wastes," *Waste Age,* 28 (1970).
3. Wilson, D. G. and O. E. Smith. "Mechanized Reclamation from Municipal Solid Waste," paper supplied by MIT, Mechanical Engineering Department, Cambridge, Mass.
4. Abrahams, J. H., Jr. "Glass Containers as a Factor in Municipal Solid Waste Disposal," *Glass Industry,* 216 (1970).
5. Boettcher, R. A. "Air Classification for Reclamation of Solid Wastes," ASME Paper 69-WA/PID-9 (1969).
6. Jensen, J. W., J. L. Holman, and J. B. Stephenson. "Future Economics from Plastic of Waste Origin," paper Presented at the First Texas Solid Waste Management Conference, Houston, Texas (May 28, 1971).
7. Kenahan, C. B. and P. M. Sullivan. "Let's Not Overlook Salvage," *APWA Reporter,* 5 (March, 1967).

8. Grubbs, M. R. and K. Ivey. "Recovery of Plastic from Urban Refuse by Electrodynamic Techniques," paper given at SPE RETEC (October, 1972).

9. McKenna, M. L. "Sorting Technology and its Application to Materials Management," ASME Paper 69-MH-14 (1969).

10. Warner, A. J., C. H. Parker, and B. Baum. "Solid Waste Management of Plastics," *Solid Waste Disposal Methods*. Research Report to Manufacturing Chemists Association, p. A-69 (December, 1970).

PART II

INCINERATION

CHAPTER 3

INCINERATION PRINCIPLES

GENERAL CONSIDERATIONS

In the ideal incineration process,[1] hydrocarbon compounds of the combustible refuse combine chemically with the oxygen of the air to form carbon dioxide and water, and leave the minerals and the metals as solid residue. This chemical reaction, called oxidation, releases high energy which can sterilize the residue, destroy odorous compounds in the refuse, and convert the water into vapor which, together with the carbon dioxide, becomes an acceptable and invisible part of the atmosphere.

As in any other chemical process, if the constituents are not intimately mixed in the proper proportions, and if they are not sustained at the proper temperature for the proper length of time, the reaction will be incomplete and undesirable products and effects may result. An uncontrolled rubbish fire which causes smoke, airborne particulate matter, odors and putrescible residue is an example of incomplete combustion.

For nearly one hundred years engineers and technicians have been developing the art of incineration of solid wastes in equipment designed for this purpose. It has become evident that if the goal of efficient, sanitary and acceptable refuse volume reduction is to be achieved through combustion, the entire incineration process, from the receipt of the refuse to the discharge of clean gas and sanitary residue, must be considered as an integrated system.

In sorting, mixing or preparing the refuse, some communities attempt to segregate refuse during collections, while others take anything that will go into a packer truck. In still others, furniture and large metal objects like refrigerators, stoves, bedsprings and bicycles are brought to the incinerator. Sometimes there are commercial wastes such as large packing crates, or industrial wastes such as large wooden pallets, or rubber tires, or spoiled batches of foodstuffs. Occasionally, noncombustibles like concrete slabs, china sinks or rolls of fence wire appear. Obviously, certain of these items should be kept out of an incinerator, while others must be treated in some special way to get them into and through the incinerator without causing damage.[1] A requirement of the feeding system is to supply a controlled flow of refuse to the furnaces with minimum interference with air supply for combustion, and with maximum protection against flashbacks of fire or gases through the charging opening.

The most popular feeding system by far is comprised of a below-grade storage bin and a travelling crane with a grab bucket which lifts and carries the refuse high above the furnace and releases it into a funnel-shaped hopper which leads to a chute that allows the refuse to slide into the furnace under the action of gravity. A few smaller incineration plants use monorail hoists in which a single fixed overhead *rail* supports a *trolley* which can travel the length of the rail on wheels. The trolley contains electric motors and brakes, and drums of steel cable which suspend, raise and lower a *clamshell*-type bucket or grapple of usually 2 to 4 cubic yard capacity over the refuse storage bin. (The halves of a bucket may be visualized as two cupped hands with the fingers of each tight together, while a grapple would look like two hands with the fingers spread apart in a claw configuration.) With monorail cranes the bin is narrow with steep sloping sides to make the refuse fall under the line of travel accessible to the bucket, and the furnace feed hoppers must also be in this line of travel to accept the refuse from the crane.

The larger incinerators have bridge cranes in which two parallel overhead rails mutually support a cross structure, or bridge, on wheels, so the bridge can travel the length of the rails. The bridge, in turn, supports a trolley which suspends and operates the bucket or grapple as described above. The bucket of the bridge crane can reach any point in the area between the support rails and can therefore handle refuse in wide bins and reach furnace charging hoppers in more locations. Usually, the crane is operated by a man in a cab mounted on the bridge or trolley.

Good cranes are costly to operate because of the sophisticated controls, the severe duty and the need for reliability, since a crane stoppage shuts down the entire plant. They also require special provisions in the buildings which house them, such as strong mountings for the rails, headroom and side clearance for the trolleys and bridges, a heavy duty, well-protected electrical power source to the trolley (either the *third rail* type or festooned retractable cables), and occasionally storage space for standby bridges, trolleys and buckets.

Tractors with bulldozing blades and lifting buckets are simpler and less costly than travelling cranes, but they can only be used to feed a furnace hopper where the refuse does not have to be taken out of the below-grade bin and lifted above the furnace.

Continuous chain, bucket or belt-type conveyors are rarely used in feeding incinerator furnaces. This is not because of inability to transport the waste material, but is perhaps due to the dismal prospect of a disabled conveyor buried 50 feet below 500 tons of mixed refuse, or to the inability of a single conveyor to stack and mix refuse like a crane or tractor. In at least one small incinerator, a tractor is used to push refuse from a tipping and sorting floor into a shallow floor trench onto an apron type metal conveyor which carries the refuse up an incline into the furnace.[1]

The most frequently used charging equipment consists of a smooth metal-lined chute extending several feet down from the *throat* of the hopper into the furnace, and terminating above one end of the stoker or hearth. The resultant column of refuse forms an air seal, and the lower end of the column of refuse is exposed to the heat of the furnace for drying and ignition. The stoking action starts the ignited refuse on its way through the furnace, and new refuse from the column replaces it.

Most materials that truly burn, *i.e.*, combine with oxygen, must first be converted to their gaseous form by heating. Much of the material in mixed refuse has moisture on it (surface moisture) or absorbed in it (inherent moisture), and any heat that is applied first turns this water to steam but leaves the burnable material too cool to volatilize and ignite until most of the water has been driven off. To perform the drying function and to prevent smothering of a going fire with undried and non-combustible material, most furnaces have some provision for exposing newly charged material to radiant heat energy and hot gases to drive off and absorb the material. As previously mentioned, these provisions take the form of exposure at the bottom of a feed chute, on a drying stoker or hearth, or brief suspension in hot gas as the refuse falls onto the stoker through the opening in the roof of an incinerator furnace.

After the moisture has been driven off, the heat radiated to the refuse by the hot gases and hot surfaces, and conveyed to the refuse by the motion of hot gases, increases the temperature of the refuse until the hydrocarbon compounds vaporize, decompose and begin to combine with oxygen. This is the ignition process which starts the burning. In most furnaces, the original ignition is done by a match, a pilot oil or gas burner, and thereafter the burning refuse ignites the incoming refuse. Subtle design features such as positioning of grates with respect to heat-reflecting walls, or guiding of flaming gases over the incoming refuse, are employed to ensure auto-ignition. In some furnaces, to ensure ignition when there are unusually wet loads, auxiliary gas or oil burners are positioned to play their flames on the incoming refuse.

The burning process consists of the continued heating and vaporizing of the elements that will combine with oxygen until only inert minerals and metals remain as ash. Actually, there are many more complex reactions during burning, and most of the metals actually oxidize to some extent, as do other elements like sulfur and even nitrogen in the intense heat of the furnace, but these are minor effects and would only confuse the general picture if discussed here.

The heart of the incineration system (Figure 3.1) is the furnace, which consists of a chamber to contain the gaseous reaction, a stoker to transport the refuse through the furnace and agitate the refuse to expose new surface to oxygen and heat, an air supply to furnish oxygen for combustion, and a pressure differential (draft) to cause the gaseous products of combustion to flow out of the chamber.[1]

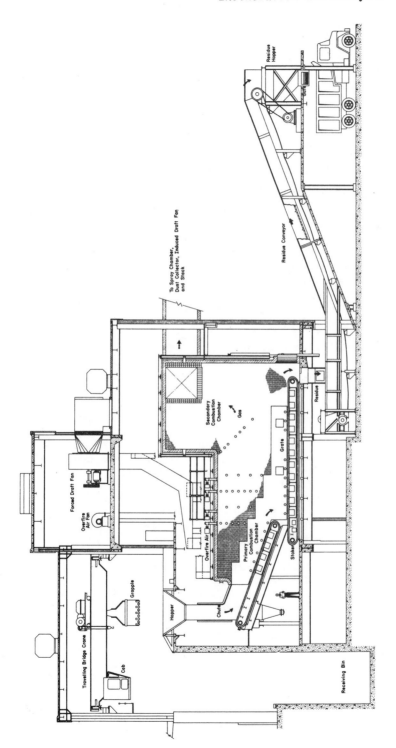

Figure 3.1. Elements of a municipal refuse incineration system.[1]

HISTORY AND EARLY INCINERATOR
DESIGN CONSIDERATIONS

In the United States

The first incinerator on record in the United
States was built on Governor's Island, New York
Harbor, in 1885. The first municipal incinerator
in America was a 30 ton per day garbage crematory
constructed at Allegheny City, Pennsylvania in 1885,
but the earliest furnace type that came into extensive use was the Engle cremator built in Des Moines,
Iowa in 1887. Other designs followed rapidly, and
by 1908 some 208 municipal incinerators had been
built in the United States. Of these, 108 were
later discontinued or abandoned. By 1921 more than
200 municipal plants were in operation, as discussed
by Stephenson.[2]
Most furnaces had a hearth onto which fresh
garbage was charged, a coal or wood burning grate
over which hot gases passed, and sometimes a second
coal or wood burning grate near the chimney. This
basic design was continued in most incinerators
built prior to World War II, although the use of
auxiliary fuel was eliminated in some of the later
plants which were designed to burn mixed refuse
rather than straight garbage.
The circular furnace was especially popular with
designers in the 1940s and 1950s. Up to that time
it had been incorporated in some 138 plants with
individual furnace design capacities of up to 175
tons per day. Following the popularity of this type
of design, mechanical stoking and continuous feed
were developed. The first of the continuous feed
furnaces was the rotary kiln plant built in Atlanta,
Georgia in 1941, and the first plant using a circular,
mechanically-stoked unit was installed in Orlando,
Florida in 1942.
The first continuous feed, rocking grate incinerator in this country was the 250-ton unit that went
into operation in Greenwich, Connecticut in 1963.
In the past 25 to 30 years incinerator design has
progressed from manually stoked to mechanically
stoked, batch feed operations to mechanically stoked,
continuous or semi-continuous feed. Today a batch
feed design is a rare exception, usually found only
in a small unit. While there have been many recent
changes and improvements in U.S. incinerator technology, none has been so universally adopted as the
continuous feed principle.[2]

Design of furnaces as simple rectangular boxes enclosing the grates or stokers is changing--many designs now provide a sloping roof arch at the discharge end of the furnace. Advantages include improved gas flow pattern and greater reflected heat on the refuse bed. With the development of the continuous feed furnace, American designers generally abandoned the principle of passing hot furnace gas over newly charged refuse before entering the combustion chamber to aid in drying and igniting the fresh material. Instead, designers locate the gas exits at the discharge end of the furnace. On the other hand, many modern European furnaces have their gas exits near the feed end, with the benefit of drying and igniting fresh refuse being one of the considerations in its location. Most recent European furnaces also have roofs much lower than those in American units in order to obtain maximum benefit from reflected heat.[2]

World War II marked a definite turning point in all aspects of incinerator design and operation, and the advances in technology since the war have been many times those of the preceding seventy years. Today, incineration is coming more and more to the fore as one of the suitable means of refuse disposal available to many municipalities. Because of the demand that the operation be efficient and nuisance-free, improvements in incinerator design, construction, and operation are being made more frequently now than at any time in the past.

The well recognized change in composition of *average* municipal refuse has probably affected incinerator design requirements more than any other single factor. Whereas the refuse mixture previously had been predominantly garbage, it has in the past 25 years changed to the point where garbage now may comprise as little as 5-10% of the total. As a result, mixed refuse is much more highly combustible, and the problem facing designers is usually one of coping with high instead of low temperatures.[2]

In Europe

Refuse Systems

As discussed by Hotti and Tanner,[3] there were three established systems of refuse destruction in Europe 25 years ago:

- Cell furnaces--the system of Heenan and Froude, Worcester
- Rotary kilns--the system of Volund, Copehhagen
- Shaft furnaces--the system of Didier, Stettin

Cell furnaces consist of a number of combustion chambers arranged side-by-side. They contain flat grates in which the refuse burns in batches. Although the cells employ mechanically driven feed gates, grates and clinker removers, they require hand stoking.

The *rotary kiln* marked a great advance, dispensing with manual stoking. The refuse, dried on stepped feed grates, continuously drops onto the firing grate where it is set afire. If it burns out completely in the rotating drum, the red-hot clinker drops from the end of the drum into a water-filled channel and is washed away.

The *shaft furnace*, in terms of process technology, occupies a position between the cell furnace and a rotary kiln. Charging may be continuous, but clinker removal is intermittent.

Design Principles

At the time, the design of the above furnace types utilized two basic principles:[3]

- The combustion chamber must not be cooled. Designers thought that cooling would not permit reaching an adequately high combustion temperature.
- There must be a separate burn-out element, either a clinker grate (cell furnace), a drum (rotary kiln), or the lowest section of the pit (shaft furnace). Without such an element, it was impossible to achieve a satisfactory burn-out of the clinker.

The most modern design at the time was the Volund rotary kiln. It was the only system to provide continuous processing and to dispense with manual work-- and it could be built in fairly large units.

Radiation Drying

European experience clearly showed that the weak points of the rotary furnace lay in the two-section gas path and in the drum as the burn-out element. This was the point at which all further development had to start. As a result, designers abandoned the division of the gas stream and passed *all* combustion

gases across the refuse lying on the pre-drying
grate. This system left both the drying and the
firing of the refuse to radiation alone. The drum
gave way to a vertical shaft (clinker generator) as
the burn-out element.

The first plant of this type went into service
in Berne in 1954, and the theory was fully confirmed
by practice. Both the drying and firing of the
refuse was reliably insured by radiation, and a good
burn-out of the clinker was achieved. In 1957,
Von Roll renewed the furnaces and boilers of the
refuse destruction plant at Borsigstrasse in Hamburg.
Built in the mid-thirties, it had suffered badly
during World War II. Despite this, the engineers
decided to retain the building, but this caused
difficulties with respect to the dimensions of the
Berne-type furnaces. They solved the problem by
arranging the pre-drying and main grates one behind
the other and working in the same direction, instead
of one above the other and in opposite directions
as in Berne. This step simplified the process.
The awkward intermediate levels of the Berne-type
furnace disappeared.[3]

Combustion Chambers

All of these plants, however, complied with
traditional principles. They had *uncooled* combustion
chambers and definite burn-out elements in the form
of clinker generators; the boilers included in all
of the plants for the purpose of utilizing the combustion heat were straight waste-heat boilers. A
new feature (compared to the Berne type) was the
integrated combustion chamber sized to keep the
heat load below 11,000 BTU per cubic foot. The fuel
bed of a refuse fire is very uneven, unlike that of
a coal fire. It exhibits a haphazard pattern of
dark patches and bright lance-like flames. These
produce locally high gas velocities which entrain
light refuse constituents such as paper. The large
combustion chamber prevents such pieces from drifting
into the boiler tubes. It also permits the complete
burn-out of the gases and the reduction of the
pressure wave arising from flashing or bursting
materials, such as films, plastics, explosives and
sealed containers.[3]

Grate Materials

The experience gained in the Berne, Brussels and Lausanne plants proved the value of highly heat-resistant grate material where the refuse contains an extremely high proportion of coal (Brussels), and where plastic waste raised the calorific value of the refuse (BASF). The first furnace of this type, embodying these grate materials and other features, was placed in service in Hamburg in 1959, and several such Von Roll plants started operating in quick succession thereafter.

Von Roll Experiences

The designers and operators of these Von Roll plants had to solve several problems, common to all such plants:

- Adapting the firing to extremes of refuse compositions;
- Influencing the calorific value of the refuse by mixing various kinds of refuse;
- Controlling the temperature of the combustion chamber;
- Adjusting the boiler size to the grate capacity;
- Working out the best boiler arrangement in order to minimize soilage and ensure long intervals between cleanings.

Tests and experience solved the first two problems. It was found possible to incinerate refuse without using any booster fuel, if it met these criteria:[3]

Water content:	50% max
Ash content:	60% max
Combustible content:	25% min

Temperature Control

Obtaining good combustion required absolute control of the combustion chamber temperature. Researchers found the temperature range narrow. It must not drop below the odor limit (about 1380° F) nor exceed the clinker softening point (about 1920° F). An operator employs any of three techniques to adjust it. He can inject air, preheat the undergrate draft or feed back flue gases. This

enables him to balance momentary variations in
refuse compositions and calorific values.

Boiler Considerations

Matching the boiler size to the grate capacity
implies a knowledge of the limits of grate load and
the efficiency of the firing system. Numerous tests
have shown that the grate can be loaded within the
limits of 50 to 100 pounds per square foot and
150,000 to 300,000 BTU per square foot, depending
on the composition of the refuse. The correlation
between firing efficiency and grate load should be
obvious--as the latter rises, the former drops and
vice versa. In the Hamburg type of destructor, as
the calorific value of the refuse rises, temperatures in the uncooled combustion chamber can no
longer be controlled by air injection and flue gas
feedback alone. Thus, designers have adopted
radiation-heating surfaces for refuse-firing systems.
There is strong precedent for this design principle
since it is used for firing systems involving the
higher grade fuels.
Boiler design and arrangement have also under-
gone considerable development. The trend was toward
adjustable radiation heating surfaces in the combustion chamber and toward cooling the ash as far as
possible before the gases entered the contact heating
surface area. As an example of boiler design--in
the Ludwigshafen plant--the boiler comprises the
radiation chamber, a succeeding downward empty flue
and the third flue, in which the contact heating
surfaces (evaporators, economizers and air pre-
heaters) are located. The superheater is arranged
either as a radiation unit, or as a contact unit in
the third flue, or both. Generally, combustion can
be maintained even when the calorific value of the
refuse is abnormally low; steam production can be
regulated and it is possible to operate the boiler
at full capacity, irrespective of the quantity and
calorific value--from low to very high, as with
high charges of plastics--of the refuse supplied.[3]

Factors Other than Design

In considering factors of incineration other
than design, the central incinerator burns waste
from many sources, contrasting with on-site

incinerators which burn material at the point of
origin, such as at an industrial plant, a store or
other commercial establishment or a large apartment
building. Generally, incineration reduces the
volume of material to be handled in a nuisance-free
manner, but it is not a complete method of disposal
except in the case of materials which are 100%
combustible and which constitute 100% of the incinerator charge in a given run. Even here destruction
is not complete, there being the matter of fly-ash
and other airborne types of solids to be accounted
for in the process. Industrial incineration is
much more likely to involve 100% combustible
material at 100% of a given run's charge. An
example would be plastic products, such as polyolefin, polystyrene and polyvinyl chloride. In
municipal incineration, which consumes by far the
greatest amount of the country's refuse, the chance
for such conditions to occur is considered extremely
remote.

ADVANTAGES OF MUNICIPAL INCINERATION

One of the chief advantages of municipal incineration is that it requires much less land than does
the landfill method--the incinerator itself can be
situated on a relatively small parcel of land.[4]
Other advantages include:
• Residue disposal requires less land than is
required in landfilling raw refuse, a significant
contribution to conserving land resources.
• An incinerator can be located much closer to
the area it serves than can a landfill.
• An incineration plant produces a residue that
contains almost no organic matter and is less
nuisance than unburned refuse. However, it is often
misstated that residue from an incinerator is *a
priori* sterile--in fact the residue produced by an
incinerator often has to be covered like raw refuse
because of the minute quantities of organic matter
remaining in the residue.
• An incinerator, well-designed and in proper
operating condition, will burn most combustibles to
an ash and can even reduce the bulk of some noncombustible components of a mixed refuse, but very
large, bulky objects or those which cause excessive
smoke are still problems for disposal by incineration.
• Some flexibility exists in the incinerator for
handling varying amounts of refuse. By modifying the

hours of operation, the number of days of operation, and the grate speed to regulate the time material remains in the combustion chamber and thus its passage through the furnaces, the incinerator plant is capable of handling a wide range of refuse volumes created by the population area it serves.

DISADVANTAGES OF MUNICIPAL INCINERATION[4]

• The incinerator plant is relatively expensive in capital cost as well as in operating cost. The initial cost, depending upon the type and size of unit constructed, will vary from $4,000 to $10,000 per ton or more of daily rated capacity. The operating cost can range from $5.00 to $9.00 per ton or more of daily rated capacity, including residue disposal and amortization.
• Skilled employees are required to operate, repair and maintain an incinerator. Such personnel are greater in number and are generally higher paid than personnel operating a landfill.
• Maintenance and repair costs are also high because of the type of equipment involved in the furnaces. Equipment can be damaged by wire, fusible metals, or abrasives entering the furnace with the refuse. It is often difficult to obtain the best sites for incinerator location because refuse disposal operations are not acceptable to many people; therefore, incinerator locations are frequently confined to industrial areas in the hope that nuisance complaints from residents nearby will be minimized.
• Incineration does not complete the job of disposal of a given community's waste; residue and fly-ash must be transported to a burial site.

REFERENCES

1. Pearl, D. R. "Technical Economic Study of Solid Waste Disposal Needs and Practices," Volume IV, Part 4, Public Health Service. Publication No. 1886, Office of Solid Waste Management Programs, Environmental Protection Agency.
2. Stephenson, J. W. "Incineration Today and Tomorrow," *Waste Age*, 2 (May, 1970).
3. Hotti, G. and R. Tanner. "How European Engineers Design Incinerators," *American City* (June, 1969).

4. American Public Works Association. *Municipal Refuse Disposal*, 3rd edition (Interstate Printers and Publishers, 1970), Chapter 5.

CHAPTER 4

BASIC DESIGN CRITERIA FOR INCINERATION
OF MUNICIPAL REFUSE

CHARACTERISTICS OF MUNICIPAL REFUSE
TO BE CONSUMED

To determine the anticipated pollutants from incineration of refuse, the composition of the refuse should be established. Municipal refuse is composed of many complex compounds, each varying in amount from small traces to large percentages of the total weight of refuse charged.[1] The basis for satisfactory incinerator design and operation is the correct analysis of the waste to be destroyed and the selection of proper equipment to best destroy that particular waste.

A sample of municipal refuse contains many major classifications of waste materials as shown in Table 4.1.[1] Typical refuse material analyses used to determine the combustion products in the gas leaving the furnace are shown. The percentage of certain items can vary seasonally. As seen from the data, major classifications have been grouped according to chemical composition. Each of the groups can be further subdivided.

There are six classes of waste, according to data presented by the American Public Works Association (APWA)[2] as shown in Table 4.2. The classification of incinerators is presented by the Incinerator Institute of America (IIA):[3]

Class I--Portable, packaged, completely assembled, direct fed incinerators having not over 5 cu ft storage capacity, or 25 lb per hour burning rate, suitable for Type 2 waste.

Class IA--Portable, packaged or job assembled, direct fed incinerators having 5 cu ft to 15 cu ft primary chamber volume; or a burning rate of 25 lb

37

Table 4.1

Typical Refuse Analysis[1]

| Refuse Type | Percent of Total Refuse by weight, dry bases | |
	Individual	Total
Plants and Grass		
Leaves	2.52	
Grass and other	5.04	7.56
Paper Products		
Newspapers	10.33	
Corrugated	23.92	
Magazines	7.48	
Wrapping Paper	6.12	
Other Paper Products	8.16	56.01
Food Wastes		
Meat and Fats	5.04	
Vegetable and Citrus	4.20	9.24
Metals		7.53
Glass and Ceramics		8.50
Plastics, Leather, Rubber		4.34
Miscellaneous		
Wood Products and Rags	3.36	
Balance, all other	3.46	6.82
		100.00

per hour up to, but not including, 100 lb per hour of Type O, Type 1 or Type 2 waste; or a burning rate of 25 lb per hour up to, but not including, 75 lb per hour of Type 3 waste.

Class II--Flue-fed, single chamber incinerators with more than 2 sq ft of burning area suitable for Type 2 waste. This type of incinerator is served by one vertical flue functioning both as a chute for charging waste and for carrying the products of combustion to atmosphere. This type of incinerator has been installed in apartment houses or multiple dwellings.

Table 4.2

Classification of Wastes to be Incinerated[2]

Type 0--A mixture of highly combustible waste such as paper, cardboard cartons, wood boxes, and floor sweepings from commercial and industrial activities. The mixture contains up to 10% by weight of plastic bags, coated paper, laminated paper, treated corrugated cardboard, oily rags, and plastic or rubber scraps.
   This type of waste contains 10% moisture and 5% noncombustible solids and has a heating value of 8,500 BTU per pound as fired.

Type 1--A mixture of combustible waste such as paper, cardboard cartons, wood scrap, foliage, and floor sweepings from domestic, commercial, and industrial activities. The mixture contains up to 20% by weight of restaurant waste, but contains little or no treated paper, plastic, or rubber wastes.
   This type of waste contains 25% moisture and 10% incombustible solids and has a heating value of 6,500 BTU per pound as fired.

Type 2--An approximately even mixture of rubbish and garbage by weight. This type of waste, common to apartment and residential occupancy, consists of up to 50% moisture and 7% incombustible solids and has a heating value of 4,300 BTU per pound as fired.

Type 3--Garbage such as animal and vegetable wastes from restaurants, hotels, hospitals, markets, and similar installations.
   This type of waste contains up to 70% moisture and up to 5% incombustible solids and has a heating value of 2,500 BTU per pound as fired.

Type 4--Human and animal remains, such as organs, carcasses, and solid organic wastes from hospitals, laboratories, slaughterhouses, animal pounds, and similar sources, consisting of up to 85% moisture and 5% incombustible solids, and having a heating value as low as 1,000 BTU per pound as fired.

Type 5--Gaseous, liquid, or semiliquid by-product waste, such as tar, paint, solvent, sludge, and fumes from industrial operations. BTU values must be determined for the individual materials to be destroyed.

Type 6--Solid by-product waste, such as rubber, plastic, and wood waste from industrial operations. BTU values must be determined for the individual materials to be destroyed.

Class IIA--Chute-fed multiple chamber incinerators for apartment buildings, with more than 2 sq ft of burning area, suitable for Type 1 or Type 2 waste (not recommended for industrial installations). This type of incinerator is served by a vertical chute for charging wastes from two or more floors above the incinerator and a separate flue for carrying the products of combustion to atmosphere.

Class III--Direct fed incinerators with a burning rate of 100 lb per hour and over, suitable for Type 0, Type 1, or Type 2 waste.

Class IV--Direct fed incinerators with a burning rate of 75 lb per hour or over, suitable for Type 3 waste.

Class V--Municipal incinerators suitable for Type 0, Type 1, Type 2 or Type 3 wastes, or a combination of all four wastes, and rated in tons per hour or tons per 24 hours.

Class VI--Crematory and pathological incinerators, suitable for Type 4 waste.

Class VII--Incinerators designed for specific by-product wastes, Type 5 or Type 6.

Solid by-product wastes such as rubber, plastics, wood waste and other solid materials from industrial operations are in Type 6 waste. Table 4.2, "Classification of Wastes to be Incinerated," shows the waste class, components, BTU value and other data. BTU values are usually set on individual materials to be destroyed. Of the seven classes of incinerators set up by IIA, Class VII is prescribed for Type 6 solid wastes. For Type 0 trash, which would include scrap plastic from municipally collected waste, the municipal incinerator Class V is prescribed. Class VII incinerators are described in the IIA standards[3] as retort types, designed for specific by-product waste. The Class V incinerators are the large municipal ones--presumably their design will permit combustion of anything which will burn satisfactorily in the Class VII retort or in-line type of industrial or commercial unit.

COMPOSITIONAL DATA OF REFUSE

Municipal Refuse

A comprehensive listing on average municipal refuse composition and analysis is shown in Table 4.3, which is from studies made by Purdue University.[4]

Table 4.3

Composition and Analysis of Average Municipal Refuse
From Studies Made by Purdue University[4]

Component	Percent of All Refuse by Weight	Moisture (percent by weight)	Analysis percent dry weight		Calorific Value (BTU/lb)
			Volatile Matter	Noncom-bustibles[a]	
Rubbish, 64%					
Paper	42.0	10.2	84.6	6.0	7572
Wood	2.4	20.0	84.9	1.0	8613
Grass	4.0	65.0	--	6.8	7693
Brush	1.5	40.0	--	8.3	7900
Greens	1.5	62.0	70.3	13.0	7077
Leaves	5.0	50.0	--	8.2	7096
Leather	0.3	10.0	76.2	10.1	8850
Rubber	0.6	1.2	85.0	10.0	11330
Plastics	0.7	2.0	--	10.2	14368
Oils, paints	0.8	0.0	--	16.3	13400
Linoleum	0.1	2.1	65.8	27.4	8310
Rags	0.6	10.0	93.6	2.5	7652
Street sweepings	3.0	20.0	67.4	25.0	6000
Dirt	1.0	3.2	21.2	72.3	3790
Unclassified	0.5	4.0	--	62.5	3000
Food Wastes, 12%					
Garbage	10.0	72.0	53.3	16.0	8484
Fats	2.0	0.0	--	0	16700
Noncombustibles, 24%					
Metals	8.0	3.0	0.5	99.0	124
Glass and ceramics	6.0	2.0	0.4	99.3	65
Ashes	10.0	10.0	3.0	70.2	4172
Composite Refuse, as Received					
All refuse	100.0	20.7	--	24.9	6203

[a]Ash, metal, glass, and ceramics.

The table lists "rubbish" by class of material, food wastes and non-combustibles. It also shows the percent of each component in the total of "all refuse," the moisture content and the BTU values per pound. Table 4.4 shows data for packaging wastes commonly found in municipal refuse.[5] Elemental analyses, inerts present and BTU values are given.

The "normal" chemical pollutants in a municipal incinerator effluent gas may be grouped into general categories: inorganic gases and particulate matter; and organic gases and particulate matter.[1]

Inorganic gases--The inorganic gases consist primarily of oxides of sulfur, oxides of nitrogen, possible precursors of other inorganic acids (*e.g.*, HCl) and ammonia. The inorganic particulates consist primarily of the oxides of such metals as aluminum, silicon, potassium, calcium, iron, titanium, zinc, sodium and magnesium. Formation of complex oxides of aluminum and silicon is also possible.

Organic gases--The organic gases and particulate matter consist primarily of fatty acids, esters, aldehydes, hydrocarbons and oxides of carbon. Most are present as gases although the fatty acids may also be present as particulates. Table 4.5 shows data.

For a given refuse composition, the quantity of particulates appears to increase with increased grate action and underfire airflow and to decrease with improved combustion. The distribution of particle size varies with combustion efficiency, underfire airflow and character of refuse. Furnaces operated in excess of design capacity show a larger weight of particulate matter per pound of flue gas. Size analysis of this particulate indicates low percentages of particulates smaller than 10 microns. Furnaces operated with low underfire air rates and at less than rated capacity show large percentages of particulates smaller than 10 microns.[1]

In general, large percentages (up to 50%) of the particulate matter can be of a combustible nature. Of this combustible particulate, amounts up to 50% may remain as acetone-solubles on analysis.

At least two approaches could be used in the determination of the possible pollution products that would be present in the gaseous effluent from

Table 4.4

Analyses of Packaging Wastes, Dry Basis %[5]

Waste	Carbon	Hydrogen	Oxygen	Nitrogen	Sulfur	Inerts[a]	BTU/lb
Corrugated paper boxes	43.73	5.70	44.93	0.09	0.21	5.34	7,429
Brown paper	44.90	6.08	47.84	0.00	0.11	1.07	7,706
Paper food cartons	44.74	6.10	41.92	0.15	0.16	6.93	7,730
Waxed milk cartons	59.18	9.25	30.13	0.12	0.10	1.22	11,732
Plastic coated paper	45.30	6.17	45.50	0.18	0.08	2.77	7,703
Newspaper (packing)	49.14	6.10	43.03	0.05	0.16	1.52	8,480
Polyethylene	85.6	14.4	-	-	-	-	19,950
Vinyl[b]	47.1	5.9	18.6 (Chlorine 28.4%)	-	-	-	8,830
Plastic film[c]	67.21	9.72	15.82	0.46	0.07	6.72	13,846
Textiles	46.19	6.41	41.85	2.18	0.20	3.17	8,036
Softwood, pine	52.55	6.08	40.90	0.25	0.10	0.12[d]	9,150
Hardwood, oak	49.49	6.62	43.39	0.25	0.10	0.15[d]	8,682
Glass bottles	0.52	0.07	0.36	0.03	0.00	99.02	84[e]
Metal cans	4.54	0.63	4.28	0.05	0.01	90.49	742[e]

[a]Ash, glass, metal.
[b]Vinyl chloride -vinyl acetate copolymer.
[c]Mixed, from municipal refuse, contaminated with food waste.
[d]Without nails, screws, etc.
[e]BTU in labels, coatings, and remains of contents of containers.

Table 4.5

Anticipated Organic Substances in
the Incinerator Effluent Gas[1]

Type	Wood, Wood Products, Plants and Grass, Food Waste	Rubber	Plastics	Vary With Excess Air
Organic acids:				
Formic	x			
Acetic	x			
Palmitic	x			
Stearic	x			
Oleic	x			
Palmitoleic	x			
Esters:				
Methyl acetate	x			
Ethyl acetate	x			
Ethyl stearate	x			
Aldehydes:				
Acetaldehyde	x		x	x
Formaldehyde	x		x	x
CO	x		x	x
CO_2	x		x	x
Polynuclear hydrocarbons		x		
Halogenated hydrocarbons[a]			x	

[a] Also from pressurized can chemicals.

municipal incinerators. One approach would be to
review the data and results of the actual sampling
and chemical composition of the refuse using published data, if available. The second approach is
a theoretical one in that the effluent constituents
are determined by estimating the combustion products
based on the chemical composition of the refuse.[1]

Nature of Municipal Refuse

Using concepts based on the assemblage of much direct data, Niessen and Chansky[6] studied the nature of refuse. Data obtained by OSWMP from a number of independent studies were reduced and statistically analyzed, as shown in Tables 4.6 and 4.7. These data exclude the categories of "Yard Wastes" and "Miscellaneous."

Table 4.6

Estimate of National Annual Average Composition of Municipal Refuse[6]

Component	Samples Utilized[a]	Mean Weight Percent	Mean (100% Total)	Standard Deviation S(X)	Confidence Limits (95%)
Glass	23	9.7	9.9	4.37	1.89
Metal	23	10.0	10.2	2.18	0.93
Paper	23	50.3	51.6	11.67	5.04
Plastics	9	1.4	1.4	0.96	0.74
Leather, rubber	9	1.9	1.9	1.62	1.25
Textiles	17	2.6	2.7	1.80	0.93
Wood	22	2.9	3.0	2.39	1.06
Food wastes	23	<u>18.8</u>	19.3	10.95	4.73
		97.6			

[a]Several data sets were not presented in a form suitable for extracting the weight fractions of all of the above refuse components.

Table 4.7

*Estimated Annual Average (% by weight)
of 1968 Municipal Refuse on "As-Discarded" Basis*[6]

Category	Unseasonal State (e.g., Florida)	Semiseasonal State (e.g., Alabama)	Seasonal State (e.g., Massachusetts)
Glass	7.6	8.1	8.8
Metal	7.5	8.1	8.7
Paper	32.6	35.1	38.2
Plastics	1.0	1.1	1.1
Leather, rubber	1.3	1.4	1.5
Textiles	1.8	1.9	2.0
Wood	2.3	2.4	2.7
Food wastes	18.2	19.5	21.1
Yard wastes	26.1	20.7	14.1
Miscellaneous	1.6	1.7	1.8

In projecting the physical and chemical nature of refuse, Tables 4.8 and 4.9 are pertinent. Table 4.8 shows expected refuse compositions for three climatic regions, and Table 4.9 shows projected refuse properties and statistics for the same three climatic regions. The authors[6] indicate that it should be recognized that these calculated values represent best estimates of the average refuse properties in future years. As such, they are useful in showing trends in various parameters affecting incinerator performance and solid-waste load. For the period 1968 to 1980, plastics are projected to increase in the solid-waste load from about 1.0% to about 2.7%. If operating problems are associated with incineration of these waste components, current study may prevent more serious problems during the next decade.

This work points up the importance of knowing for any method of solid waste disposal, but particularly when municipal incineration is the method, the types of refuse to be consumed and their compositional data. Two of the conclusions offered by the authors of *The Nature of Refuse*[6] are significant:

Table 4.8

Projected Refuse Compositions (% by weight) on "As-Discarded" Basis[6]

Refuse Category	1968			1970			1975			1980		
	Seasonal	Semi-seasonal	Non-seasonal	Seasonal	Semi-seasonal	Non-seasonal	Seasonal	Semi-seasonal	Non-seasonal	Seasonal	Semi-seasonal	Non-seasonal
Glass	8.8	8.1	7.6	9.1	8.4	7.9	9.9	9.2	8.7	10.3	9.6	9.0
Metal	8.7	8.1	7.5	8.8	8.2	7.6	9.0	8.4	7.8	9.4	8.7	8.1
Paper	38.2	35.1	32.6	39.1	35.8	33.5	40.8	37.6	35.2	41.5	38.4	36.1
Plastics	1.1	1.1	1.0	1.3	1.3	1.1	1.9	1.8	1.7	2.8	2.7	2.5
Leather, rubber	1.5	1.4	1.3	1.5	1.4	1.3	1.5	1.4	1.3	1.5	1.4	1.3
Textiles	2.0	1.9	1.8	2.0	1.9	1.8	2.1	2.0	1.9	2.1	2.0	1.9
Wood	2.7	2.4	2.3	2.5	2.3	2.2	2.2	2.0	1.9	2.0	1.8	1.7
Food wastes	21.1	19.5	18.2	20.2	18.7	17.4	17.9	16.6	15.5	16.2	15.0	14.1
Miscellaneous	1.8	1.7	1.6	1.7	1.6	1.5	1.5	1.4	1.3	1.4	1.3	1.2
Yard wastes	14.1	20.7	26.1	13.8	20.4	25.7	13.2	19.6	24.7	12.9	19.2	24.1
Total	100.0	100.0	100.0	100.0	100.0	100.0	100.0	100.0	100.0	100.0	100.0	100.0

Table 4.9
Projected Refuse Properties and Statistics[6]

Refuse Properties and Statistics	1968			1970			1975			1980		
	Seasonal	Semi-seasonal	Non-seasonal	Seasonal	Semi-seasonal	Non-seasonal	Seasonal	Semi-seasonal	Non-seasonal	Seasonal	Semi-seasonal	Non-seasonal
Heating value (HHV, BTU/lb)	4582	4505	4449	4628	4550	4493	4719	4640	4582	4811	4730	4627
Percent moisture	25.9	27.8	29.3	25.2	27.1	28.6	23.4	25.3	26.9	22.1	24.0	25.7
Percent volatile carbon	19.8	19.4	19.1	19.9	19.5	19.2	20.4	20.0	19.7	20.8	20.4	20.1
Percent ash content	21.8	20.3	19.1	22.1	20.7	19.5	22.9	21.5	20.3	23.5	22.0	20.3
Percent ash (excluding glass, metals)	5.5	5.3	5.1	5.5	5.2	5.1	5.3	5.1	4.9	5.2	5.0	4.9
Per capita growth multiplier	1.0	1.0	1.0	1.05	1.05	1.05	1.19	1.18	1.18	1.32	1.32	1.31
National population growth multiplier	1.0	1.0	1.0	1.02	1.02	1.02	1.07	1.07	1.07	1.13	1.13	1.13
Total waste load multiplier	1.0	1.0	1.0	1.07	1.07	1.07	1.27	1.26	1.26	1.49	1.49	1.48
Per capita heat-rate multiplier (BTU/person/day)	1.0	1.0	1.0	1.06	1.06	1.06	1.23	1.22	1.22	1.39	1.39	1.36
Total heat-rate multiplier (BTU/day)	1.0	1.0	1.0	1.08	1.08	1.08	1.32	1.31	1.31	1.57	1.57	1.54

- Incinerator problems caused by low bulk density, high heating value, and high volatile content in refuse will increase in the future.
- Over the next thirty years, the quantity of refuse collected by municipalities will increase by a factor of 2.5 on a per capita basis and by a factor of almost 4 on a total quantity basis. This disposal burden should serve to spur the installation of new incinerator capacity in most urban centers.

In more recent work, Niessen and Alsobrook[7] show a statistical evaluation of national average composition estimates (Table 4.10); an estimated average municipal refuse composition (by weight) for 1970 (Table 4.11); the percent moisture in refuse on an "as-discarded" and "as-fired" basis (Table 4.12); and projected average generated refuse composition, heating value and quantity for the periods of 1970-2000 (Table 4.13).

Table 4.10

Statistical Evaluation of National Annual Average Composition Estimate[7]

Component[a]	Samples Utilized	Mean Weight Percent	Mean (100% Total)	Standard Deviation	95% Confidence Limits[b]
Food wastes	38	19.7	19.6	9.08	2.89
Glass	37	10.1	10.1	4.03	1.30
Metal	37	9.9	9.9	2.03	0.65
Paper	38	50.8	50.6	11.04	3.51
Plastics	10	1.70	1.62	1.09	0.69
Leather & Rubber	10	1.77	1.68	1.59	0.98
Plastics, Leather & Rubber (Combined)	32	3.3	3.3	1.29	0.45
Textiles	33	3.0	3.0	1.96	0.67
Wood	37	3.5	3.5	4.98	1.60
Total		100.3	100.0		

[a]Excluding yard-wastes and miscellaneous categories.
[b]We assumed that the data are as in a "normal" distribution.
[c]Note that in many of the data sets, these components are not reported separately.

Table 4.11

Estimated Average of 1970 Municipal Refuse Composition (weight percent) on an "As-Discarded" Basis[7]

Category	Summer	Fall	Winter[a]	Spring
Paper	31.0	39.9	42.2	36.5
Yard wastes	27.1	6.2	0.4	14.4
Food wastes	17.7	22.7	24.1	20.8
Glass	7.5	9.6	10.2	8.8
Metal	7.0	9.1	9.7	8.2
Wood	2.6	3.4	3.6	3.1
Textiles	1.8	2.5	2.7	2.2
Leather and rubber	1.1	1.4	1.5	1.2
Plastics	1.1	1.2	1.4	1.1
Miscellaneous	3.1	4.0	4.2	3.7
Total	100.0	100.0	100.0	100.0

[a]The refuse composition in winter for southern states is similar to that shown for fall.

Table 4.12

Percent Moisture in Refuse on an "As-Discarded" and "As-Fired" Basis[7]

Component	As-Fired	As-Discarded
Food wastes	63.6	70.0
Yard wastes	37.9	55.3
Miscellaneous	3.0	2.0
Glass	3.0	2.0
Metal	6.6	2.0
Paper	24.3	8.0
Plastics	13.8	2.0
Leather and rubber	13.8	2.0
Textiles	23.8	10.0
Wood	15.4	15.0

Note: As it is mixed together with other refuse materials, discarded solid waste may either lose or absorb moisture prior to incineration.[6]

Table 4.13

Projected Average Generated Refuse Composition,
Heating Value and Quantity, 1970-2000[7]

	1970	1975	1980	1990	2000
Composition:					
(weight %, as discarded)					
Paper	37.4	39.2	40.1	43.4	48.0
Yard wastes	13.9	13.3	12.9	12.3	11.9
Food wastes	20.0	17.8	16.1	14.0	12.1
Glass	9.0	9.9	10.2	9.5	8.1
Metal	8.4	8.6	8.9	8.6	7.1
Wood	3.1	2.7	2.4	2.0	1.6
Textiles	2.2	2.3	2.3	2.7	3.1
Leather and rubber	1.2	1.2	1.2	1.2	1.3
Plastics	1.4	2.1	3.0	3.9	4.7
Miscellaneous	3.4	3.0	2.7	2.4	2.1
(weight %, as burned)					
Moisture	25.1	23.3	22.0	20.5	19.9
Volatile carbon	19.6	20.1	20.6	21.8	23.4
Total ash	22.7	23.4	23.9	22.8	20.1
Ash (excluding glass and metal)	6.5	6.2	6.1	6.0	6.0
Relative Heating Value and Quantity:[a]					
Heating Value (BTU/lb), as fired	1.00	1.02	1.04	1.09	1.17
Heating Value (BTU/lb), dry basis	1.00	1.00	1.00	1.06	1.09
National Population	1.00	1.05	1.10	1.31	1.51
Per Capita Refuse Generation (lb/person/day)	1.00	1.13	1.26	1.44	1.66
Per Capita Refuse Heat Content (BTU/person/day)	1.00	1.15	1.31	1.57	1.94
Total Generated Refuse Quantity (lb)	1.00	1.19	1.38	1.89	2.51
Total Refuse Heat Content (BTU)	1.00	1.23	1.44	2.05	2.93

[a]Ratio relative to 1970 value.

MUNICIPAL INCINERATORS

Incinerators recently designed for municipal use are fed refuse continuously throughout the burning period. The ability to receive fuel (refuse) in this manner eliminates many of the problems associated with municipal incinerators of batch feed design. Among the advantages of continuously fed incinerators are large furnace capacity, excellent control, and near uniform furnace temperatures which reduce thermal damage to components. Two classes of continuous feed furnaces are (1) refractory-lined and (2) water wall. These two types are covered in the following discussion together with common modifications.[8]

Refractory-Lined Furnace

The refractory-lined furnace, commonly used for refuse incineration in American practice, is a proven system and is the most economical type of unit in sizes under 250 tons per day. Above this limit, increased maintenance and operating costs begin to offset the cost of adding other small units.

Refractory replacement is a costly item for this type of furnace. Refractory is damaged by overheating, slagging, chemical reactions, erosion of the furnace lining by the materials in contact with the brick and by heating and cooling of the furnace. Wide temperature variations in the furnace produce refractory damage through expansion, contraction and through moisture. When the unit is allowed to heat and cool, stresses are created in the brick and furnace frame. These stresses cause movement of the furnace lining, which in turn can chip or break the bricks. Condensation of moisture also occurs when the furnace is cooled. Moisture, which is absorbed by the refractory lining, turns to steam when the furnace is brought back to operating temperature. The rapid expansion of gas in forming steam creates a pressure within the brickwork which spalls off the face of the lining. Replacing refractory usually means a furnace is out of service for two weeks or more.

Reported evidence indicates that relining is typically required every two or three years. The cost of relining will range from $20 to $30 per square foot--more than $100,000 for a complete relining of a 300 ton-per-day furnace. In addition

to the major relining projects, minor repairs to
brickwork are often required. The cost of these
repairs varies widely, depending upon the extent of
damage, but the square-foot cost cited is appropriate.
Repair and replacement of furnace refractory account
for substantial portions of the annual cost of
operating this class of incinerator.

The brick-lined furnace relies on air for
cooling. This air must be supplied in large quantities to prevent overheating of the refractory
lining and excessive slagging of the walls. This
excess air joins the combustion gases and must be
passed through the air pollution control equipment.
The larger volume of gas being treated for particulate matter removal increases the cost of air
pollution control devices.

The refractory-lined furnace does not require
licensed personnel for operation. Unlicensed
personnel can become qualified as efficient incinerator operators and skilled in maintaining the
plant in its optimum condition. The savings in
labor and operating expenses inherent with the
refractory furnace must be weighed against the
higher maintenance cost associated with this type
of unit. Operating personnel is an important consideration of incinerator cost *vs.*efficiency. Some
operators and superintendents are inclined to
operate understaffed, both in quality and quantity
and the result is often a dirty, poorly maintained
plant and grounds plus sloppy operation of
equipment.[8]

Water Wall Furnaces

Water-filled tubes form the furnace wall in this
type of furnace.[8] The water-wall furnace for
municipal refuse incineration has found application
in European and South American installations. There
has been a water wall incinerator in operation in
Norfolk, Virginia, for nearly five years. Other water
wall furnace constructions are maintained in Chicago
and New York.

Water wall furnaces can be constructed for a
large range of capacities. The minimum practical
size of a water wall is about 600 tons per day. In
multi-unit incineration plants using large furnaces,
the ability to deliver refuse (fuel) to the plant
will become an important consideration.

The walls of the water-cooled furnace consist
of a series of tubes connected by fins or welded
directly together. Heat is transferred from the
burning refuse through the wall to the water in the
tubes. The water-filled tubes form a cool wall in
contact with the flame and hot gases, preventing
the accumulation of slag.

Water cooling reduces the use of excess air.
Excess air in water-wall furnaces can be held to
approximately 50 to 60%, although provision for
supplying 100% excess air is advisable. The volume
of air passing through the furnace has a direct
effect upon the size of air pollution control devices needed. A secondary advantage of restricting
excess air quantities is the introduction of less
moisture in the form of humidity into the gas
stream.

Moisture can be a problem in water-wall furnaces.
The moisture may combine with gases formed by the
combustion of plastics and certain other materials
found in refuse. If the moisture and gas combine,
they can form an acid capable of damaging the walls
and boiler tubes. Inspection of the Norfolk incinerator after nearly a year of operation revealed
little corrosion of the boiler tubes. The formation
of corrosive bodies can be minimized by using supplemental heat during the periods the furnace is
out of service.[8]

Tube replacement in water-wall furnaces is,
according to European experience, an infrequent
maintenance problem--such periods occur only once
in 5 to 7 years. There has not yet been enough
experience with water-wall refuse incinerators in
this country to provide an accurate forecast of wall
maintenance costs.

Water-wall furnaces require specialized and
careful treatment of the water used in the boiler.
Among the impurities to be removed are scale and
sludge-forming materials, soluble salts, oil, and
dissolved or releasable gases. Such removal may be
achieved by the same methods used to treat boiler
water in power plants.

The water-wall furnace requires skilled personnel for efficient operation. A water-wall
furnace is a boiler plant, and the installation
should employ the same qualified personnel found
on the floor of any steam-generating plant.

Recent studies have promoted water-wall furnaces
on the basis of recovery or sale of steam generated.
The effort to conserve this valuable by-product is

a laudable one, but, unless a ready market is available or can be developed, it is an impractical consideration.[8]

FURNACE MODIFICATIONS

In modifications of furnaces, many attempts have been made to improve performance. These attempts have resulted in two major modifications, namely, waste heat boilers and rotary kilns.

Waste Heat Boilers

Waste heat boilers can be installed in the furnace to cool combustion gases without introducing large amounts of excess air or moisture. Incinerator gases are cooled from 1800-2000°F to 400-600°F before discharge to the atmosphere. Cooling the gas is done to reduce gas volumes, to reduce damage to furnace components and to improve fan efficiency. Gas cooling is an important part of incinerator operation.

In general practice gas is cooled by dilution with excess air and/or water spray. Both methods add to the volume of gas to be handled by the furnace components and to the moisture content of that gas. Both methods remain popular because they require little additional capital expenditure.

The heat passed to the boiler water must be dissipated or explosions will result. Heat is con-verted to steam and may be used for heating, auxiliary power, to drive steam-driven equipment or for other purposes. In-plant use of steam will consume about 10 to 15% of the total steam produced. Excess steam should be condensed and recirculated to conserve boiler feed water. Condensation takes place in cooling towers. Forced air cooling towers mounted on the roof of the incinerator are satis-factory for condensing surplus steam.

Water wall furnaces permit the use of smaller capacity air pollution control equipment to meet existing codes with high efficiency and should provide easy accommodation for restrictions in emission standards when set.[8]

Rotary Kiln Modification

Rotary kiln modification of the refractory furnace provides a chamber in which hot gases pass

over the residue from the grates (Figure 4.1). The
kiln rotates slowly on its long axis, tumbling the
material being burned. The residue leaving the kiln
contains less putrescible material than residue from
standard grates. Some rotary kiln furnaces are of
European design and are licensed for manufacture and
sale in the United States. These units are produced
currently in the 250-ton/day size, although manufacturers are considering building larger units.
While past construction couples rotary kilns with
refractory furnaces only, it seems technically
feasible to couple a refractory kiln with a water
wall furnace. However, such a hybrid could well
have most of the disadvantages of both types of
furnaces and few of the advantages.[8]

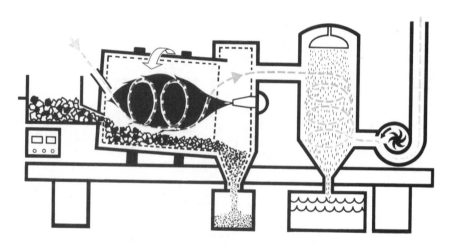

*Figure 4.1. Rotary kiln. (Courtesy of Bartlett-Snow,
Cleveland, Ohio)*

The faults of refractory and water wall furnaces
are recognized throughout the industry. Refractory
furnaces are a proven form of refuse incineration
in the United States and elsewhere, whereas there
has as yet been little experience with water wall
incinerators in this country. Nevertheless, it is
believed that the potential of the water wall may
well be superior to that of the refractory wall
type. Further, it is believed that the problems,
particularly those of corrosion of walls, can be
overcome and that the water wall furnace has more
to offer in the way of advanced technology than
does the refractory wall furnace.[8]

When operating flexibility is required, all burning systems should be as versatile as possible in the types of materials that can be handled. In refuse incineration, there are three forms of flow-through incineration systems:

1. *The grate type furnace*--this includes such grate systems as rocking, travelling, reciprocating, reverse reciprocating (Martin), and drum-type (Dusseldorf).
2. *The rotary hearth type of furnace*--the furnace is often built with several hearths, one over the other. It may have either a fixed hearth with rotating rabble arms or a rotating hearth with fixed rabble arms.
3. *The rotary kiln furnace*--burning is done on the refractory hearth which rotates around an inclined axis.

All three of these forms may be used in the design of either municipal or industrial incinerators.

GRATE SYSTEMS

In the United States, there are four principal types of flat-bed stokers in use in municipal incinerators.[9]

• The *traveling grate stoker* (Figures 4.2 and 4.3) is essentially a moving chain belt carried on sprockets and covered with separated small metal pieces called keys. The entire top surface can act as a grate while moving through the furnace, yet can flex over the sprocket wheels at the end of the furnace, return under the furnace and reenter the furnace over a sprocket wheel at the front. The sprockets drive this chain conveyor and are in turn driven by electric motors at slow speed.

• The *reciprocating grate stoker* (Figures 4.2 and 4.4) is a bed of bars or plates arranged so that alternate pieces, or rows of pieces, reciprocate slowly in a horizontal sliding mode and act to push the refuse along the stoker surface. These bars or plates are driven through links by electric motors or hydraulic cylinders.

• The *rocking grate stoker* (Figure 4.2) is a bed of bars or plates on axles. By rocking the axles in a coordinated manner, the refuse is lifted and advanced along the surface of the grate. This

Figure 4.2. Mechanical stoking equipment. (Courtesy of Plibrico Co. Commercial Booklet Municipal Waste Disposal.)

Figure 4.3. Traveling grate stoker. (Courtesy of Illinois Stoker Co., Alton, Illinois)

Figure 4.4.   *Reciprocating grate incinerator stoker.*   *(Courtesy
of Detroit Stoker Co., Monroe, Michigan)*

type of stoker is actuated by linkages driven by
hydraulic or electric motors.

  • The *circular grate* is commonly used in conjunction with a central rotating cone grate with extended
rabble arms that agitate the fuel bed.

  • Although much more widely used in Europe,
where it was designed and first built and used,
there are several municipal incinerators in the
United States fired by *rotary kilns*.   Such installations are made by licensees of the European designer.
This type of firing mechanism is really a combination
furnace and stoker[9] and is very effective in gently
tumbling the burning refuse until complete combustion
is achieved.   The kiln is a large metal cylinder with
its axis horizontal or slightly inclined.   It is
lined with firebrick and mounted on rollers so that
electric motors can slowly rotate it about its
horizontal axis.   As used in municipal incinerators,
the refuse is first passed over flat-bed type drying

and ignition grates in a furnace and then, when most moisture and volatile constituents have been driven off, the burning residue is fed into the kiln for the final burn-out. In such an arrangement, the volatiles driven off in the ignition chamber are sent through a passage above the rotating kiln and join the hot gas effluent from the kiln in a secondary combustion chamber where combustion of the gases and airborne particulate matter is carried out.

In contrast to the grates used in the United States, several species grates have been designed and developed in Europe to facilitate the burning of refuse. In regard to choice of grate system, Eberhardt[10] gives ten elements on which to base judgment:

- Adaptability of the fire to handle a great mass of refuse and residue.
- Adaptability of the refractory to handle wide radiation effects.
- Controllable air quantity, temperature and excess air.
- Adjustable retention time and variations according to the fuel.
- Adjustable height of layer.
- Controllable stabilizing heat supply.
- Controllable cooling of the residue.
- Controllable flue gas temperature before contacting the radiation heating surface.
- Observability of the fire layer and of the fire gases.
- Functioning ability of refuse firing due to technical design:
  - Prevention of reignition.
  - Positive conveyance of the refuse mass.
  - Serviceability and replaceability of wearing parts.
  - Measuring and control system.

Eberhardt[10] describes the reciprocating or feeding grate as follows:

In the *reciprocating grate* (Figure 4.4), single grate stages are arranged one upon the other. Fuel bed movement results from the direction of grate reciprocating movement; moreover, the stoking of the fire is affected by this movement. A loosening of the material to be burned, in the case of a very deep fuel layer, is quite limited. The required burnout values are obtained by using a slag generator or a

rotating drum. In the case of wide grates for high
throughput, an effort to attain burnout is arranged
through use of another grate. The grate load is
relatively low and the undergrate pressure high.
It is sometimes possible to open paper bales by
this grate arrangement because only at the points
of transfer is there a circulation. The grate arts
are moved by a common drive; in the case of failure
of one grate link, the entire fuel bed has to be
discharged.

Three European grate types are described by
Rogus:[11]

Drum Grate System--the Dusseldorf Type. This
is composed of a series of cylinders or drums placed
on a 30% downward slope. Each drum is 5 feet in
diameter and 10 feet long. They rotate in the discharge direction at adjustable peripheral speeds
varying from a maximum of 50 feet per hour, for the
first cylinder, to a minimum of 16 feet per hour in
the end cylinder. The surface of each drum is built
up of serrated bars of commercial gray iron cast in
the form of arced segments and keyed to a structural
steel spider. The slow, varied speed rotation of
each drum subjects the refuse to a mild continuous
tumbling and agitation. Each drum rests separately
over individual steel compartments for the purpose
.of zoning the input of underfire air and the siftings
discharge. The combustion of the burnable refuse is
nearly complete; the discharge of the siftings is
automatic and little dust is entrained in the exhaust
gases.

Martin Reverse Action System. This is an
adaptation, by Josef Martin of Munich, of lignite
burning grates to the handling and burning of
municipal refuse. The Martin unit is a stepped-
down grate inclined at about 30% slope towards the
discharge end. The grate surface is made of heavy
serrated cast iron bars anchored by keys into a
structural frame. These keys push uphill in a
reciprocating action against the downward flow of
the refuse. At every stroke a portion of the
burning material is pushed against and under the
still unignited refuse. In addition, a short
relative motion takes place between adjoining bars,
thereby freeing any metals, wires or clinkers that
may have lodged in the intervening air spaces. The
thickness of the refuse bed is controlled by a
reciprocating pan-type feeding gate at the forward

end, and by a variable speed clinker drum at the discharge end. Grate burning rates are about 50% higher than those used in the United States, ranging between 90 and 95 pounds per square foot per hour. A characteristic of the Martin system is its short but wide configuration. Underfire air is zoned, enabling appropriate distribution. The siftings are pneumatically discharged into the residue compartment.

Von Roll System. This is a product of the Von Roll Company of Zurich (Figure 4.5). The latest Von Roll design (1966 or earlier) transfers refuse from the charging hopper via a vibrating inclined hopper -- controlling the rate and assuring positive feed of the refuse through a vertical chute -- onto a 20% slope drying stoker. After ignition, the refuse is dropped vertically about 5 feet onto the second (30% slope) burning stoker. Finally it is dropped another 5 feet onto a third (33%) sloped stoker to complete the burnout. The hot residue is discharged into a water trough equipped with a flight conveyor or scraper. The third grate replaces the former "clinker" generator since the latter was not fully successful, particularly in the higher capacity plants. Grates are of the reciprocating stepped-down pallet type with the underfire zoned and the siftings discharged directly into the residue trough below.

Some other types of grates found in European incinerators are mentioned by Blach:[12]

Tip Grate. This is a modified design of the well-known coal-fired Huber stoker, in which the tip elements have been turned from backward moving to forward moving. The alternate raising and lowering of the grate sections gives the refuse a turning and waving movement down the grate so that new material is brought into contact with the combustion air and radiation heat can facilitate an intensive combustion in all of the combustion layer. The grate can be divided into lengthwise sections making it possible to attain very large grate widths. As is the case with the Martin grate, the angle of inclination must be rather wide in order to secure the right movement of the refuse.

Volund Forward Pushing Step Grate. This grate is made up of several sections, each separated by a vertical grate transition bar. The ratio of size between the individual grate section and grate

Figure 4.5. Model of European incineration plant. (Courtesy
of Von Roll Ltd., Berne, Switzerland)

transitions is determined by the composition of the refuse. The individual grate section consists of sections placed lengthwise and laid up with an inclination of 18 to 25°. Alternate sections are movable, the others being fixed. Each section is built up of a welded through-grate bar on which a number of grate blocks of specially alloyed cast iron are fitted. The blocks are in turn fitted with loose grate bars of cast iron. The movable sections are driven hydraulically by a transverse driving shaft under the grate. The movement forward is to raise slowly, go forward, then to lower slowly and go backward. The backward stroke is the reverse, *i.e.*, to lower and move backward, then raise and go forward. The first grate section acts as a feeding and predrying grate. Ignition and the initial part of the combustion take place at the first transition and on the second grate. The final combustion and burnout take place on the third grate, and calcining and cooling of the clinkers begin at the last part of the third grate and in the subsequent clinker chute. The movable grate sections give a lifting and turning motion in the under part of the refuse layer so that the combustion air can get to all parts of the layer. At the transition bars there is a supplementary turning, mixing, and air supply.

REFERENCES

1. Day and Zimmerman Associates. "Special Studies for In- cinerators in Washington, D.C.," Public Health Service Publication No. 1748, Office of Solid Waste Management Programs, Environmental Protection Agency.
2. American Public Works Association. *Municipal Refuse Disposal*, 3rd Edition (Interstate Printers and Publishers, Inc., 1970).
3. *Standards of Incinerator Institute of America--IIA* (1969).
4. Corey, R. C. *Principles and Practices of Incineration.* (J. Wiley and Sons), p. 7.
5. Kaiser, E. R. "Incineration of Packaging Wastes with Minimal Air Pollution," *Proceedings, First National Conference on Packaging Wastes* (September, 1969), p. 181.
6. Niessen, W. R. and S. H. Chansky. "The Nature of Refuse," paper prepared for 1970 National Incinerator Conference (ASME) at Cincinnati, Ohio (May 17-20, 1970).
7. Niessen, W. R. and A. F. Alsobrook. "Municipal and Industrial Refuse: Composition and Rates," paper prepared for 1972 National Incinerator Conference (ASME) at New York (June 4-7, 1972).

8. Jones and Henry Engineers Ltd. "Proposal for a Refuse Disposal System in Oakland County, Michigan," Public Health Service Publication No. 1960, Office of Solid Waste Management, Environmental Protection Agency (1970).
9. Warner, A. J., C. H. Parker, and B. Baum. "Solid Waste Management of Plastics," Research Study for Manufacturing Chemists Association (1970).
10. Eberhardt, H. "European Practices in Refuse and Sewage Sludge Disposal by Incineration," paper prepared for 1966 National Incinerator Conference (ASME) at New York (May 1-4, 1966).
11. Rogus, C. A. "An Appraisal of Refuse Incineration in Western Europe," paper prepared for 1966 National Incinerator Conference (ASME) at New York (May 1-4, 1966).
12. Blach, E. *Plants for the Incineration of Refuse.* Varme No. 3, English Translation (1968).

CHAPTER 5

SITE SELECTION AND UTILITIES REQUIREMENTS

MUNICIPAL INCINERATOR SITE SELECTION

A proper incinerator location enhances acceptance by the public, results in economical waste collection, facilitates efficient and economic incinerator operation, promotes a pleasing appearance and minimizes housekeeping and maintenance.[1]
Factors important to design, but generally not of concern to the public, are foundation conditions, topography, availability and immediate location of all utilities required, building restrictions, drainage and meteorologic conditions. Topography and meteorological conditions must be considered in the design of the incinerator. A flat site is apt to require a ramp for access to the tipping floor, whereas a hillside site can provide access at various ground levels. Topography can also ease or hinder the dispersion of gases and particulates by the local atmosphere. Stack height determination requires consideration of topography and legal restrictions such as Federal Aeronautics Agency regulations, local building regulations and zoning. As in the development of any industrial operation site, effective drainage of surface waters must be an integral part of design. The site should not be selected in an area subject to flooding unless the facility can be protected and access remains available during high water.
Locations central to the load from the area served usually result in the lowest collection costs. Topography and soil conditions have a distinct influence on construction costs. Hillside sites can reduce construction costs by permitting use of designs where the refuse flow roughly parallels the natural ground surface. This results in less

excavation and simpler structural design compared
with a multistoried plant on flat ground. Subsurface
soils and ground water will have a profound effect
on foundation design and underground structures.

Availability of public utilities may be a
governing factor in site selection since electricity,
gas, water supply, sewage disposal and process water
disposal are essential to the incineration process.
Fuel such as gas or oil may be required at some
installations. Communications facilities must be
available for fire and safety control and for coordinating operations. A municipal incinerator requires
certain utility services which include: electricity
for power and lighting; potable water for plant
personnel and suitable process water for spraying,
heating, quenching, cooling and fire fighting;
telephone service; sewage systems for handling
process waste and sewage, and storm sewers for
drainage; fuel for heating, hot water, auxiliary
heat for incineration and possible laboratory use.
Each of these utilities supplied to the incinerator
site must be metered and distributed safely and
efficiently to all points of usage. With increasing
incinerator capacities and with increasing use of
sophisticated equipment and devices, more utility
services will be required. The cost of providing
these utilities depends on the plant design and
mode of operation and may reasonably be expected
to range from $0.10 to as much as $1.00 per ton of
waste processed.[1]

With few exceptions, utilization of incinerator
waste heat to generate electric power is not practiced in the United States -- the electric power to
run the incineration operation is obtained from
other sources. Power requirements vary with the
degree of mechanization and the use of equipment.
Some common examples of equipment requiring electricity are induced draft fans, forced-air fans,
pumps, cranes, hoists, pollution control devices
and grate-drive mechanisms. Allowances for future
electric needs should be included in planning and
sizing the electrical distribution systems. For
some facilities, electric power can cost as much as
$0.75 per ton of waste incinerated.

The quality of the water required for incinerator
operation will depend on its use. The total amount
of water required may vary from 350 to 2,000 gallons
per ton of waste incinerated depending on design and
operation. Incinerator operation requires a dependable water supply. Water for air pollution control

equipment, for gas cooling, and for dust control sprays need not be potable but should be free from suspended materials. Water used in the incineration processes will increase in temperature, change in chemical characteristics, and will acquire solids. Treatment may be required before such waters are discharged. In reuse and recycling, treatment should be effective in preventing clogging, erosion, and corrosion of equipment.[1]

Fuels may be required for plant processes, including building and water heating, and for auxiliary fuel. The choice of fuel for these purposes will depend on availability and cost. The need is determined by local conditions and incinerator design.

Where it is possible to have some choice in selecting a site, several factors should be kept in mind:[1]

- Isolation with respect to residential areas.
- Distance, in the direction of prevailing winds, of the populated areas from the site.
- Provision for wastewater treatment before discharge to a stream.
- Availability of water supply.
- Availability of power supply.
- Accessibility by transportation equipment.
- On-site ash disposal, if possible.
- Minimal refuse and residue hauling distances.
- Proximity to plant areas in position to use waste heat.
- Cost of property if not already owned.

Some other considerations influencing selection of a site include:[1]

- Amount of rehandling of waste materials at both the collecting area and the incineration area.
- Expense of handling wastes requiring corrosion-resistant equipment and/or heated storage.
- Handling of reactive liquid residues which require segregation.
- Speed of delivery of materials which will deteriorate and produce offensive odors if stored for lengthy periods.
- Consideration of fire and toxicity hazards if waste materials are allowed to accumulate before transport to incineration area.
- Provision of economical loading conditions at the collection points and unloading conditions at the disposal point.

Meteorological conditions are a strong consideration in site selection as the significance of the
air pollution problem is closely related to geographical location and terrain. Air movement in the
geographical area which surrounds a particular
proposed incineration plant can be obtained from
U.S. Weather Bureau data. Of particular concern
are data on wind speed, direction and duration.
Inversion conditions which prevail over a significant
percentage of time may create the most stringent
requirements for emissions resulting from incineration. The federal government has shown an increasing
awareness of the air pollution problem, with increasing activity by federal regulatory agencies in
this area, and it is expected that this trend will
continue. Therefore, it is essential that a site
be chosen, if possible, in the most advantageous
location from a dispersion standpoint, and that the
design of the burning facilities be flexible enough
to accommodate efficient gas-cleaning devices, if
these were not originally installed on the burning
equipment.

REFERENCE

1. DeMarco, J., D. J. Keller, J. Leckman, and J. L. Newton.
   "Incinerator Guidelines," Public Health Service Publication
   No. 2012, Office of Solid Waste Management Programs,
   Environmental Protection Agency (1969).

CHAPTER 6

INCINERATOR CAPITAL AND OPERATIONAL COSTS

MUNICIPAL DESIGNS

National Averages on Costs

As pointed out by DeMarco, Keller, Leckman and Newton,[1] incinerator costs are divided between those related to ownership and those related to operation. Ownership costs derive from the capital costs of financing the incinerator construction. Operating costs include direct and indirect costs of operation and maintenance of the plants.

The capital costs of 170 municipal incinerators were obtained from data provided in the U.S. Public Health Service National Survey of Community Solid Waste Practices. The capital costs are reported as the 1966 estimated replacement costs and include the costs of buildings, facilities and engineering, but not land. These data indicate an average capital cost of municipal incinerators at $6,150 per ton per day design capacity. (Example: for 200 tons of such capacity, the capital cost would be $1,230,000). Sixty-two percent of the incinerators studied cost less than the average. Fifteen of the 170 plants studied reported capital costs in excess of $11,000 per ton per day. The highest capital cost, reported by only one plant, was $30,000 per ton per day.

Some of the major equipment cost items included in capital costs are scales, cranes, furnaces, blowers, air pollution control devices, process waste treatment and recyling equipment, residue removal systems, instrumentation, waste heat recovery equipment, steam distribution equipment, and flue and duct equipment. Major construction items include building, ramps, tipping area, storage pit,

71

refuse hoppers, offices, employee facilities, piping
and chimney. Miscellaneous items included under
capital cost include site preparation, excavation,
foundation preparation, road-shop equipment and
tools. Fly-ash control equipment has, in the past,
accounted for about 3% of the total cost of municipal
incinerators. To achieve particulate removal as
required by new and more stringent air pollution
control regulations, the cost of control equipment
will now range from 8 to 10% (or perhaps a little
more in the 1970-1975 period) of the total capital
cost. Capital cost components and their relative
contributions to the whole may be grouped as follows: furnaces and appurtenances, 55 to 65%;
building, 20 to 30%; air pollution control equipment,
8 to 10%; miscellaneous 7 to 13%.

National Survey Data
on Operating Costs

     The 1968 National Survey data[2] also provided
information on the costs of operating municipal
incinerators. The data gave an average cost for
operating such incinerators of $5.00 per ton of
refuse processed; 73% of the incinerators studied
had operating costs below the average of $5.00 per
ton with about 5% of those reporting showing
operating costs above $10.00 per ton. The wide
variation in operating costs resulted partly from
differences in the amount and types of pollution
control equipment, labor rates, cost of utilities,
residue disposal costs and amounts of automation.
For some incinerators, the costs of utilities,
administration and employee benefits are not included in operating cost. The costs for depreciation
and interest have commonly been given as between
$1.00 and $2.00 per ton processed. Influences here
would be factors such as utilization rate, estimated
life and interest rates. Total costs can be stated
for a time period or for a quantity incinerated;
e.g., annual cost or cost per ton. Operating cost
comparisons with previous years do not show much
change from the average of $5.00 per ton processed
shown by the National Survey.

European *vs* U.S. Costs

     Exact cost comparisons between European and
American installations are difficult because of

differences in costs of labor and materials, in the
refinements used, and in the number and type of
auxiliaries installed. However, Rogus[3] has made
some broad but dependable evaluations based on Tables
6.1 and 6.2.

    · In Table 6.1 on capital costs, the European
incinerators cited include such elements as steam
power generation, electrostatic precipitation, and
oversized-wastes processing equipment. Few of these
items are regularly incorporated in United States
plants. The table shows that unit costs per ton per
day range from $6,000 to $16,000 with a median of
about $10,000. This appears to be well in line with
the $8,000 adjusted unit cost for New York City plants.
    · On operating costs, Table 6.2 shows some
data obtained from abroad compared with that available
for New York City incinerators. Amortization costs
were excluded.

    Rogus[3] states that gross operating, maintenance,
repair and supervisory costs in Europe are from 10
to 25% lower than for New York City even though
operating and maintenance costs for a number of
revenue-producing auxiliaries are included. This
is so because if proper credit is taken for the
sale of steam, power and various salables (residue,
metals, fly ash), their net operating costs become
lower. This lower operating cost also reflects
lower unit labor costs, somewhat higher skills and
productivity and, in addition, the nature of
incinerator designs, workmanship and sturdiness of
construction.

ENGINEERING COST DATA,
SPECIFIC INSTALLATIONS

Oakland County, Michigan

    Jones and Henry Engineers Ltd.[4] in their proposal
for incinerator systems for Oakland County, Michigan
considered four systems in projecting costs. These
were: (a) water wall furnace with waste heat
boiler; (b) refractory lined furnace; (c) refractory
lined furnace with waste heat boilers; and (d) refractory lined furnace with rotary kiln.
    The comparisons were made of 300 ton-per-day
units to be operated in plants containing from one
to five such units. As seen in Table 6.3, initial

Table 6.1

Capital Costs[3]

Plant Location*	Furnaces Types, No. & Size (ton/day)	Supplementary Fuels	Water cooled walls	Refractory walls	Waste heat boilers	Special — Electr. precipitators
Vienna	Von Roll 3 @ 200	Waste oil	–	X	X	X
Munich	Martin 2 @ 660	Pulv. coal	X	–	X	X
	Martin 1 @ 1060	Pulv. coal	X	–	X	X
Dusseldorf	Drum 4 @ 250	–	X	–	X	X
Rotterdam	Martin 4 @ 385	–	X	–	X	X
Paris (Issy-les-Moulineaux)	Martin 4 @ 450	–	X	–	X	X
Lausanne	Von Roll 2 @ 200	–	–	X	X	X
Montreal	Martin or Von Roll 4 @ 300	–	X	–	X	X
N.Y.C. (6 mod. plants)	Travel' Gr. 4 @ 250	–	–	X	–	–

*All plants burn mixed, unsegregated refuse.

Table 6.1, continued

Steam gen. plant	Electric gen. plant	Oversized waste eq't.	Residue & metals salv.	Chimneys per plant	Capital Cost		Remarks
Auxiliaries					Total Cost for Plant	Unit Cost per ton per day	
X	X	-	X	1	$9,600,000	$16,000	Difficult foundations
X	X	X	X	1	21,750,000	9,100	Costs approx.
X	X	X	X	1			
X	X	X	X	1	7,500,000	7,500	Cost incl. bldg. for total of 6 furnace units
X	X	X	X	2	9,250,000	6,000	Steam for on-site only, El. power gen.: 10% for inc. bal. sold
X	X	-	X	2	20,000,000	11,100	Highly sophisticated
X	X	-	X	1	4,000,000	10,000	Handsome plant in midst of apartment area
X	-	X	X	2	12,000,000	10,000	Bids taken Nov. 2, 1965
-	-	-	-	2	8,000,000	8,000	Large subsidence chambers

Table 6.2

Operating Costs[3]

Plant*	Design Capacity per plant (ton/day)	Operating hr per wk	Operating Factor[a]	Tons Processed (tons/wkg day)	Man Power (man-hr/wkg day)	Cost (man-hr/ton)	Cost[b] (dollars/ton)
European							
A	600	168	70	420	384	0.91	—
B	1,000	168	85	850	564	0.66	—
C	1,540	126	70	980	592	0.60	$2.50 net
D	400	—	50	200	—	—	$3.50 gross $1.50 net
U.S.A. 6 Modern continuous-type plants[c]	1,000	128	80	300	576	0.72	$4.30

Note: All European incinerators equipped and operated to include steam and power generation, residue processing and metal salvage.

[a] Operating Factor = average actual production ÷ design capacity.

[b] Total cost exclusive of amortization.

[c] Average values per plant for 6 modern plants for 3 years of 300 working days each -- refuse incineration only.

Table 6.3

Typical Comparative Cost Figures for Furnace Systems[4]
(3 units @ 300 tons/day)

	Water Furnace with Waste Heat Boiler	Refractory Lined Furnace	Refractory Lined Furnace with Waste Heat Boiler	Refractory Furnace with Rotary Kiln
Incinerator Structure	$2,240,000	$2,290,000	$2,240,000	$2,290,000
Utility Construction	259,000	265,000	256,000	265,000
Furnace Components	2,933,000	903,000	2,375,000	3,253,000
Air Pollution Control Equipment	690,000	1,500,000	900,000	1,500,000
Estimated Project Cost[a]	$6,122,000	$4,958,000	$5,771,000	$7,308,000
Amortization of Capital Cost Based on 4 1/2% Annual Interest, 20-year bonds	$ 472,000	$ 380,000	$ 447,000	$ 563,000
Plant Labor	520,000	495,000	520,000	496,300
Utilities	15,000	35,000	15,000	36,500
Building Supplies, Operation & Maintenance	193,000	312,000	280,000	350,400
Estimated Annual Cost[a]	$1,200,000	$1,220,000	$1,262,000	$1,446,200

[a]Excluding cost of land, landscaping and residue disposal.

comparison of units based on capital costs showed
the refractory furnace with rotary kiln was the
most expensive type of all of the units, while
refractory-lined furnaces were the least expensive.
This is true, despite the fact that larger air pollution equipment control was included for both the
refractory furnace and the refractory furnace with
rotary kiln, in order to achieve equal stack emissions
in terms of particulate matter.

A more significant basis of comparison is that
of annual cost. When capital costs are reduced to
annual bond retirement costs, and operation, maintenance, labor and other annual costs are included
in the comparison, the advantage no longer remains
exclusively with the refractory furnace. This
situation is brought about because of the increasing
maintenance and operating costs of refractory furnaces in sizes exceeding 250 tons per day--installa-
tions with 1 to 5 furnace units consistently show
this reversal of cost.

In Table 6.3 the item "Incinerator Structure"
includes building, foundation and stack. The item
"Utility Construction" consists of the cost of
extending public utility lines to the plant and the
cost of treating wastewater from the incinerator.
Grates, refractory or water walls and arches,
scales, cranes, hoppers, fans, duct work and, where
applicable, boiler banks and auxiliaries are included under the heading "Furnace Components." The
item "Air Pollution Control Equipment" consists of
electrostatic precipitators and associated equipment.
All cost figures are for 1970.

In evaluating cost/performance of these four
types, the annual cost of operating furnaces with
waste heat boilers varies with the class of furnace.
In the case of the refractory type, the additional
cost of boiler tube maintenance and the cost of
processing feed water must be added to the already
elevated cost of operation and maintenance. Most
of these additional annual costs do not apply in
the case of water wall furnaces. Here any additional
cost would be for the added capacity of boiler water
treatment equipment and for tube repair in the
boiler bank.

A substantial added expense is incurred by providing a boiler bank and condensing equipment in a
refractory furnace instead of a conventional cooling
system. The cost of adding waste heat boiler equipment to a refractory furnace is $1500 to $2500 per
ton of rated capacity. The cost for comparable

equipment in a water wall furnace is $2500 to $3500 per ton. The cost is lower for the refractory furnace because large quantities of dilution air are used for gas cooling, a situation not encountered in the water wall furnace. With air dilution, the size of the boiler bank required to lower combustion gas temperature to a suitable range can be reduced.[4]

The refractory-lined furnace is economical in terms of initial cost for all unit sizes and where there are from one to five units per plant. The average range of capital cost for refractory incinerators is $4,000 to $7,000 per ton of rated capacity, depending on unit sizes and number of units per plant. The capital cost of constructing water wall incinerators with waste heat boilers is in the range of $4,000 to $9,000 per ton of rated daily capacity. Recent bids taken in Chicago for water wall furnaces with waste heat boilers approached $11,000 per ton of rated daily capacity. Operating and installation costs for rotary kiln modification to refractory furnaces are readily available in 250 ton-per-day units. Construction costs range from $5,000 to $9,000 per ton of rated daily capacity, adjusted to 1970 prices. Operating and maintenance costs are approximately 20% greater than for refractory furnaces of equal capacities without the kiln modification.

Refractory furnaces show a good annual cost ratio in sizes under 275 tons per day. Such furnaces should be considered for use in all sizes up to 300 tons per day. Manufacturers of water wall furnaces recommend furnace sizes for their equipment ranging from 200 to 600 tons per day per furnace. Economic evaluation indicates that this equipment is more advantageous in terms of annual cost in sizes above 300 tons per day. A practical limit to water wall equipment probably will be 500 tons per day per unit. The use of water wall equipment in sizes under 300 tons per day can be economical only if waste heat can be economically used.[4]

Washington, D.C.

Day and Zimmerman[5] studied four basic configurations of furnace, boiler and exit gas cooling for incinerators planned for Washington, D.C. Considered for their effect on performance and operating costs were: refuse composition and heating value; relative characteristics of refractory and water cooled

furnaces; specific design requirements of boiler
convection surfaces suitable for use with incinerator
furnace gases; the effect of heat recovery equipment
on flue gas volume due to spray water cooling; the
effect of steam demand requirements on incinerator
operation, and the effect of variations in refuse
moisture and furnace excess air on steam production
rates. The costs in Table 6.4 are 1968 estimates.

Cost Analysis for Seven Incinerators

Achinger and Daniels[6] gathered data from six
municipal incinerators and one pilot incinerator as
follows:

A. Built in 1966 as a 300 ton/day design
capacity unit, it consists of two refractory-lined,
multiple-chambered furnaces with inclined, modified
reciprocating grate sections followed by a stationary
grate. It is provided with a wet scrubber consisting
of 42 twelve-inch-diameter wetted columns.

B. Also built in 1966 with the same capacity
as A, also refractory-lined, multiple-chambered with
three sections of inclined rocking grates. Air
pollution control equipment consists of a wet
scrubber with flooded baffle walls.

C. A pilot plant unit built in 1967 with a
capacity of 5,000 to 6,000 pounds in ten or twelve
hours burning. The furnace consists of a conical
burner with double metal walls and fixed grates.
It includes an after burner, a centrifugal type
water scrubber and an electrostatic precipitator.

D. Built in 1965 as a 500 ton/day installation, consisting of two refractory-lined, multiple-
chambered furnaces with two sections of traveling
grate, one inclined and the other horizontal, it
includes a wet scrubber consisting of flooded baffle
walls.

E. Built in 1963 with the same capacity as
D, it is comprised of two furnaces with three
reciprocating grate sections followed by a rotary
kiln. It also has a wet scrubber with water sprays
and a baffle wall.

F. Also built in 1963 with a capacity of 600
tons/day and the same features as E.

Table 6.4

Estimated Capital Investments and Operating Costs, [a] Incinerator and Incinerator Boiler Plants, Refuse Design Capacity—4,000 Tons per Week[5]

Item	Case I Refractory Furnace	Case II Refractory Furnace with Boiler	Case III Waterwall Furnace with Boiler	Case IV Waterwall Furnace
Capital costs:				
General building construction	$ 930,000	$1,215,000	$1,030,000	$ 938,000
Equipment delivered to site	1,340,000	2,620,000	2,230,000	2,150,000
Mechanical contract	1,075,000	1,580,000	800,000	983,000
Electrical contract	238,000	322,000	237,000	232,000
Total incremental cost	$3,583,000	$5,737,000	$4,297,000	$4,303,000
Annual operating expenses:				
Operating days per week	5	7	7	5
Maintenance labor and supplies	$ 140,000	$ 161,000	$ 147,000	$ 123,000
Operating labor	326,000	495,000	495,000	365,000
City water	28,000	25,000	43,000	30,000
Auxiliary fuel	---	140,000	187,000	10,000
Electric power	161,000	184,000	174,000	161,000
Operating supplies and chemicals	1,000	2,000	3,000	1,000
Subtotal	$ 656,000	$1,007,000	$1,049,000	$ 690,000
Fixed charges on investment	$ 276,000	$ 442,000	$ 331,000	$ 331,000
Estimated total annual expense	$ 932,000	$1,449,000	$1,380,000	$1,021,000
Estimated value of steam per 1,000 pounds	---	$ 1.06	$ 0.49	---

[a] Capital investments and operating expenses include only those variables affected by plant design. They are not intended to include all costs of operation or construction at the incinerator plant.

G.   Built in 1967 of 400 tons/day capacity,
it is comprised of two refractory lined furnaces
with four sections of inclined, reciprocating grates,
followed by multiple dry cyclones and by a wet baffle
wall.

Data on these installations are given in the
following tables. Table 6.5 shows that total annual
costs, including amortization and interest, varied
from $4.02 for incinerator A to $9.36 for incinerator
G.   Incinerator C was a pilot plant, and meaningful
cost data were not available. Also shown in Table
6.5 is the adjusted projected annual cost at design
capacity--these projected costs were determined by
prorating costs that depend on the quantity of
material processed, actual *vs* design. Also adjusted
to a common reference point so that data from various
incinerators could be compared in Table 6.5 were labor
costs, plant depreciation and interest. The annual
costs are, of course, the sum of operating costs and
capital investment (financing and ownership costs).
Capital investment in the incinerators varied from
$1,800 to $8,400 per ton of design capacity as shown
below. Because Incinerator A did not have scales,
residue quench tanks, a crane or a storage pit, the
investment was obviously less.

Analysis of Capital Investment

Incinerator	Actual Cost	Adjusted Cost	Adjusted Cost/Ton
A	$ 471,659	$ 541,276	$ 1,804
B	1,848,240	2,121,040	7,070
D	3,000,000	3,564,300	7,129
E	3,321,779	4,214,341	8,429
F	2,400,000	3,044,880	5,075
G	2,530,855	2,793,052	6,983

The capital costs were adjusted to the year 1967
by the use of a construction cost index. Briefly,
this index was developed from unpublished data
developed by the American Society of Civil Engineers
and the American Society of Mechanical Engineers
which showed that the capital costs for constructing
incinerators are divided between building and equipment in a 60/40 ratio. Thus, the index for incinerators A and B, both built in 1966, is 1.0526; for D,
built in 1965, it is 1.1036; and for E and F, both
built in 1963, the index is 1.1881. The adjusted
cost per ton is then obtained by dividing the
adjusted capital cost by the design capacity of the
plant.

Table 6.5

Annual Cost Data[6]

Item	Incinerator A				Incinerator B			
	Normal Capacity		Design Capacity		Normal Capacity		Design Capacity	
	Actual	Adjusted	Projected	Adjusted	Actual	Adjusted	Projected	Adjusted
Operating costs:								
Direct labor	$ 75,184	$ 73,703	$112,776	$110,555	$197,500	$185,730	$197,500	$185,730
Utilities	17,352	17,352	35,135	34,135	20,000	20,000	25,850	25,850
Parts and supplies	12,509	12,509	24,608	24,608	32,950	32,950	42,580	42,580
Vehicle operations	1,739	1,739	3,421	3,421	7,200	7,200	9,300	9,300
External repairs	7,346	7,346	14,451	14,451	6,250	6,250	8,080	8,080
Disposal charges	700	700	1,377	1,377	2,000	2,000	2,580	2,580
Overhead	11,326	11,326	11,326	11,326	52,800	52,800	52,800	52,800
Total operating cost	126,156	124,675	202,094	199,873	318,700	306,930	338,690	326,920
Operating cost/ton	2.95	2.92	2.41	2.38	4.90	4.72	4.03	3,89
Financing & ownership costs:								
Plant depreciation	23,581	23,581	23,581	21,651	80,149	84,842	80,149	84,842
Vehicle depreciation	6,042	6,042	6,042	6,042	9,675	9,675	9,675	9,675
Interest	16,059	32,477	16,059	32,477	64,558	127,262	64,558	127,262
Total financing and ownership cost	45,682	60,170	45,682	60,170	154,382	221,779	154,382	221,779
Financing and ownership cost/ton	1.07	1.41	0.54	0.72	2.38	3.41	1.84	2.64
Total cost	171,838	184,845	247,776	260,043	472,082	528,709	492,072	548,699
Total cost/ton	4.02	4.33	2.95	3.10	7.28	8.13	5.87	6.53

Table 6.5, continued

Item	Incinerator D Normal Capacity Actual	Adjusted	Design Capacity Projected	Adjusted	Incinerator E Normal Capacity Actual	Adjusted	Design Capacity Projected	Adjusted
Operating costs:								
Direct labor	$193,138	$186,301	$193,138	$186,301	$202,407	$205,139	$202,407	$205,139
Utilities	18,000	18,000	19,301	19,301	65,260	65,260	90,418	90,418
Parts and supplies	0	0	0	0	57,332	57,332	79,433	79,433
Vehicle operations	7,670	7,670	8,225	8,225	4,188	4,188	5,802	5,802
External repairs	22,339	22,339	23,954	23,954	1,999	1,999	2,770	2,770
Disposal charges	32,232	32,232	34,562	34,562	0	0	0	0
Overhead	32,959	32,959	32,959	32,959	123,577	123,577	123,577	123,577
Total operating cost	306,338	299,501	312,139	305,302	454,763	457,495	504,407	507,139
Operating cost/ton	2.35	2.29	2.23	2.18	4.50	4.53	3.60	3.62
Financing & ownership costs:								
Plant depreciation	200,000	142,572	200,000	142,572	110,726	168,574	110,726	168,574
Vehicle depreciation	0	0	0	0	3,516	3,516	3,516	3,516
Interest	70,448	213,858	70,448	213,858	106,859	252,860	106,859	252,860
Total financing and ownership cost	270,448	356,430	270,448	356,430	221,101	424,950	221,101	424,950
Financing and ownership cost/ton	2.07	2.73	1.93	2.55	2.19	4.20	1.58	3.04
Total cost	576,786	655,931	582,587	661,732	675,864	882,445	725,508	932,089
Total cost/ton	4.42	5.02	4.16	4.73	6.69	8.73	5.18	6.66

Table 6.5, continued

Item	Incinerator F				Incinerator G			
	Normal Capacity		Design Capacity		Normal Capacity		Design Capacity	
	Actual	Adjusted	Projected	Adjusted	Actual	Adjusted	Projected	Adjusted
Operating costs:								
Direct labor	$165,684	$181,391	$165,684	$181,391	$150,949	$145,434	$160,949	$145,434
Utilities	67,632	67,632	70,500	70,500	31,952	31,952	75,777	75,777
Parts and supplies	51,540	51,540	53,725	53,725	2,700	2,700	6,403	6,403
Vehicle operations	9,600	9,600	10,007	10,007	13,968	13,968	33,127	33,127
External repairs	12,758	12,758	13,299	13,299	808	808	1,916	1,916
Disposal charges	10,364	10,364	10,803	10,803	27,720	27,720	65,741	65,741
Overhead	84,674	84,674	84,674	84,674	21,331	21,331	21,331	21,331
Total operating cost	402,252	417,959	408,692	424,399	259,428	243,913	365,244	349,729
Operating cost/ton	2.49	2.59	2.43	2.53	5.49	5.17	3.26	3.12
Financing & ownership costs:								
Plant depreciation	80,000	121,795	80,000	121,795	101,234	111,722	101,234	111,722
Vehicle depreciation	0	0	0	0	0	0	0	0
Interest	75,840	182,693	75,840	182,693	81,494	167,583	81,494	167,583
Total financing and ownership cost	155,840	304,488	155,840	304,488	182,728	279,305	182,728	279,305
Financing and ownership cost/ton	0.97	1.89	0.93	1.81	3.87	5.91	1.63	2.50
Total cost	588,092	722,447	564,532	728,887	442,156	523,218	547,972	629,034
Total cost/ton	3.46	4.48	3.36	4.34	9.36	11.08	4.89	5.62

Table 6.6

Repair and Maintenance Cost Data[6]

Item	Incinerator A		Incinerator B		Incinerator D	
	Actual	*Adjusted*	*Actual*	*Adjusted*	*Actual*	*Adjusted*
Expenditures						
Expenditure type:						
Labor	$10,442	$10,237	$29,625	$27,861	$53,590	$51,695
Parts	12,509	12,509	32,951	32,951	0	0
External charges	7,346	7,346	6,250	6,250	22,339	22,339
Overhead	1,574	1,574	7,919	7,919	9,145	9,145
Total	31,871	31,666	76,745	74,981	85,074	83,179
Allocation						
Cost center:						
Receiving and handling	$ 6,824	$ 6,780	$ 9,112	$ 8,903	$13,766	$13,460
Volume reduction	21,641	21,502	56,825	55,518	60,109	58,770
Effluent handling & treatment	3,406	3,384	10,808	10,560	11,199	10,949
Total	31,871	31,666	76,745	74,981	85,074	83,179

Table 6.6, continued

Incinerator E		Incinerator F		Incinerator G	
Actual	Adjusted	Actual	Adjusted	Actual	Adjusted

Expenditures

Incinerator E		Incinerator F		Incinerator G	
$61,335	$62,164	$31,679	$34,685	$44,003	$39,762
57,332	57,332	51,540	51,540	2,700	2,700
1,999	1,999	12,758	12,758	808	808
37,447	37,447	16,189	16,189	6,799	6,799
158,113	158,942	112,166	115,172	54,310	54,069

Allocation

Incinerator E		Incinerator F		Incinerator G	
$39,345	$39,552	$17,960	$18,442	$18,642	$17,187
85,597	86,045	77,804	79,888	17,834	16,441
33,171	33,345	16,402	16,842	17,834	16,441
158,113	158,942	112,166	115,172	54,310	50,069

Table 6.7

Operating Cost Breakdown by Cost Centers[6]

Cost Center	Incinerator A Actual	Incinerator A Adjusted	Incinerator B Actual	Incinerator B Adjusted	Incinerator D Actual	Incinerator D Adjusted
Receiving and handling:						
Direct labor	$25,062	$24,568	$79,000	$74,290	$51,942	$50,103
Utilities	0	0	3,020	3,020	12,600	12,600
Vehicle operating expense	1,564	1,564	0	0	0	0
Repairs and maintenance	6,824	6,824	9,112	9,112	13,766	13,766
Overhead	3,775	3,775	21,119	21,119	8,864	8,864
Total	37,225	36,731	112,251	107,541	87,172	85,333
Volume reduction:						
Direct labor	35,504	34,805	59,250	55,720	48,451	46,736
Utilities	8,597	8,597	7,480	7,480	2,700	2,700
Repairs and maintenance	21,641	21,641	56,826	56,826	60,108	60,108
Overhead	5,348	5,348	15,839	15,839	8,268	8,268
Total	71,090	70,391	139,395	135,865	119,527	117,812
Effluent handling and treatment:						
Direct labor	4,176	4,094	29,625	27,860	39,156	37,770
Utilities	8,755	8,755	9,500	9,500	2,700	2,700
Vehicle operating expense	175	175	7,200	7,200	7,670	7,670
Disposal charges	700	700	2,000	2,000	32,232	32,232
Repairs and maintenance	3,406	3,406	10,809	10,809	11,199	11,199
Overhead	629	629	7,920	7,920	6,682	6,682
Total	17,841	17,759	67,054	65,289	99,639	98,253
Total	126,156	124,881	318,700	308,695	306,338	301,398

Table 6.7, *continued*

Incinerator E		Incinerator F		Incinerator G	
Actual	Adjusted	Actual	Adjusted	Actual	Adjusted
$67,470	$68,381	$59,294	$64,915	$73,360	$66,288
6,964	6,964	12,715	12,715	18,720	18,720
0	0	0	0	0	0
39,345	39,345	17,960	17,960	18,642	18,642
41,192	41,192	30,302	30,302	9,081	9,081
154,971	155,882	120,271	125,892	119,803	112,731
30,667	31,081	34,226	37,471	14,260	12,885
7,724	7,724	8,251	8,251	6,193	6,193
85,597	85,597	77,804	77,804	17,834	17,834
18,725	18,725	17,492	17,492	1,819	1,819
142,713	143,127	137,773	141,018	40,106	38,731
42,935	43,515	40,485	44,323	29,325	26,498
50,572	50,572	46,666	46,666	77,039	77,039
4,188	4,188	9,600	9,600	13,968	13,968
0	0	10,364	10,364	27,720	27,720
33,171	33,171	16,402	16,402	17,834	17,834
26,213	26,213	20,691	20,691	3,632	3,632
157,079	157,659	144,208	148,046	99,519	96,691
454,763	456,668	402,252	414,956	259,428	248,153

Data from Achinger and Daniels on repair and
maintenance cost data are shown in Table 6.6.
Operating costs are analyzed in Table 6.7 to determine the relationship between labor, utility, and
repair and maintenance costs. In all cases, labor
costs were highest and utilities lowest. Labor
costs averaged 58% and far exceeded all other costs;
repair costs were also higher at some centers than
at others. In general, labor costs were the major
portion of operating costs at every facility.

Cost Analysis of East New Orleans Incinerator

In another study of incinerator costs, Heil[7]
states that in February 1969 a cost analysis of the
East New Orleans municipal incinerator was made by
the U. S. Department of Health, Education and Welfare
to determine both total and unit operating costs.
Table 6.8 shows a total annual cost of $557,878.
On the basis of a capacity of 400 tons per day, the
unit cost is $4.73 per ton. The table also shows a
total operating cost of $375,150 and a total
ownership cost of $182,728. Broken down into unit
costs, these are $3.18 per ton and $1.55 per ton,
respectively, for operation and ownership. If the
unit cost were calculated on the basis of actual
tons burned, and the plant continues to operate at
nearly 20% over design capacity, the actual unit
operating cost would be reduced to less than $4.00
per ton burned. This is an ultramodern operation
based on two 200 ton/day furnaces rather than one
400 ton/day furnace to give greater operating
flexibility. Stokers are of the automatic,
continuous-feed reciprocating grate type. Instrumentation is extremely well done and air pollution
control equipment is multicyclone. The total plant
construction cost came to $2,214,090--a unit cost
approximately $5,520 per ton daily rated capacity.
The plant became operational on final acceptance in
May 1968.

Chicago Northwest Incinerator

In an excellent coverage of the basis for design
from an engineering standpoint, Fife[8] describes the
1600 ton/day Northwest Incinerator in Chicago. This
plant is the first precipitator-equipped, refuse-
burning, water-walled furnace, steam-generating

Table 6.8

Annual Cost Breakdown[7]

Description	Cost	Totals
Direct Labor	$160,949	
Utilities	79,880	
Parts and Supplies	6,750	
Vehicle Operating Expense	34,920	
External Repair Charges	2,020	
Disposal Charges	69,300	
Overhead	21,331	
Total Operating Cost		$375,150
Depreciation	101,234	
Interest	81,494	
Total Ownership Cost		182,728
Total Annual Cost		$557,878

incinerator engineered in the United States. It is expected that its successful operation will be followed by the construction of at least four other similar plants. The Northwest Incinerator, in operation, will be the largest municipal incinerator in the Western Hemisphere.

The low bidder offered the European Martin (reverse-acting, reciprocating) stoker--the first such in the U.S. The bid on the refuse-burning system included steam system, residue handling system, stoker type, boiler auxiliaries, water-walled furnace and electrostatic precipitators. There were six separate contracts for the Northwest Incinerator:

1.	Refuse-burning system	$7,994,007
2.	Foundation and structures	2,482,818
3.	Cranes	602,600
4.	General construction and site work	5,466,919
5.	Shredders	289,688
6.	Stacks	612,000
	Total	$17,448,092

Table 6.9

Comparative Cost Data for Alternate 300-ton/day Refuse-Burning Systems[8]

System		Furnace and Gas. Cleaning Equipment	Estimated Equipment Costs[a]	
	Stoker		Installed First Cost($)	Cost/ton of Capacity ($)
Recommended	Bar & Key traveling grate (American manufacture)	Water-wall and precipitator	1,149,000	3,830
Alternate A	Reciprocating grate (American manufacture)	Water-wall and precipitator	1,009,000	3,364
Alternate A European (Martin)	Reverse acting reciprocating grate (European)	Water-wall and precipitator	1,200,000	4,000
Alternate C	Bar & Key traveling grate (American manufacture)	Refractory walls, spray chamber, and precipitator	675,785	2,253
Alternate B	Reciprocating grate (American manufacture)	Refractory walls, spray chamber, and precipitator	588,946	1,963
Alternate B European (Martin)	Reverse acting reciprocating grate (European manufacture)	Refractory walls, spray chamber, and precipitator	816,475	2,722
Alternate D	Reciprocating grate and rotary kiln (American manufacture)	Refractory walls, spray chamber, and precipitator	712,841	2,376

[a]Equipment costs include elements of one complete refuse-burning system. This includes the cost of steam-condensing equipment.

[b]Annual power cost is based on 6832 h at $0.015 kWh. Water-wall system uses steam power for induced-draft fans, boiler feed pumps and forced draft fans.

[c]Process water cost is based on 6832 h at $0.22/1,000 gal.

[d]Maintenance costs include allowances for refractories, boiler feedwater treatment, packing, lubrication, and stoker parts. Labor cost is taken as an annual charge of $8,000/year for one

Table 6.9, continued

Estimated Annual Costs				Total Fixed Charges and Operating Costs ($/yr)	Estimated Cost/ton Burned ($)	Fly-Ash Emission Gr/st ft³[f]
Power[b] ($/yr)	Process Water[c] ($/yr)	Operation and Maintenance[d] ($/yr)	Amortization[e] ($/yr)			
10,500	None	13,985	81,579	106,064	1.24	0.10
11,300	None	13,985	71,369	96,924	1.13	0.10
11,200	None	13,985	85,200	110,385	1.29	0.10
24,510	10,822	29,375	47,981	112,688	1.32	0.10
25,336	10,822	29,375	41,815	107,348	1.26	0.10
31,092	10,822	29,375	57,970	129,257	1.51	0.10
26,161	10,822	33,650	50,612	121,245	1.42	0.10

operator per unit. Cost of labor and operation is based on 1200 tons/day.

[e]Amortization includes interest at 4% and repayment of principal over a 20-year period.

[f]Grains of particulate per standard cubic foot.

NOTE: This data is not for use in construction estimating.

Changes in the installation and other factors are believed to have pushed the costs closer to $25,000,000.[8]

The data on estimated total owning and operating costs (dollar/ton burned) are:[8]

Operating costs	$3.95
Amortization[a]	3.40[b]
	7.35

Revenue from sales	
Steam	2.10
Salvaged metals	0.25
	2.35

Net cost per ton burned	$7.35
	-2.35
	$5.00

[a] Based on actual bid prices which also show that installed first cost of refuse-burning system equipment is about half of total project cost.

[b] This figure must now be raised 50%.

Table 6.9 shows full details on comparative costs of four bid systems and includes estimated equipment costs and estimated annual costs.

Fife concludes his work with the following observations:

- The effects of stringent air pollution requirements on the total plant design are significant and must be recognized by designers of large plants early in their work.
- The data show the overall effect of the solid-wastes management plan on plant sizing, reliability and construction-schedule requirements.
- Water-walled furnaces require considerable lead-time in construction.

REFERENCES

1. DeMarco, J., D. J. Keller, J. Leckman, and J. L. Newton. "Incinerator Guidelines," Public Health Service Publication No. 2012, Office of Solid Waste Management Programs, Environmental Protection Agency (1969).
2. Office of Solid Waste Management Programs, Environmental Protection Agency. "Massive Statistics," National Survey, 1968, Public Health Service Publication No. 1867 (1968).
3. Rogus, C. A. "Recent Trends of Refuse Collection and Disposal in Western Europe," paper prepared for 1966 National Incinerator Conference (ASME) at New York. (May 1-4, 1966).
4. Jones and Henry Engineers Ltd. "Proposal for a Refuse Disposal System in Oakland County, Michigan," Public Health Service Publication No. 1960, Office of Solid Waste Management Programs, Environmental Protection Agency (1970).
5. Day and Zimmerman Associates. "Special Studies for Incinerators in Washington, D.C." Public Health Service Publication No. 1748, Office of Solid Waste Management Programs, Environmental Protection Agency.
6. Achinger, W. C. and L. E. Daniels. "An Evaluation of Seven Incinerators," paper prepared for 1970 National Incinerator Conference (ASME) at Cincinnati, Ohio (May 17-20, 1970).
7. Heil, T. C. "Planning, Construction and Operation of the East New Orleans Incinerator," paper prepared for 1970 National Incinerator Conference (ASME) at Cincinnati, Ohio (May 17-20, 1970).
8. Fife, J. A. "Design of the Northwest Incinerator for the City of Chicago," paper prepared for 1970 National Incinerator Conference (ASME) at Cincinnati, Ohio (May 17-20, 1970).

CHAPTER 7

INCINERATOR INSTRUMENTATION

INSTRUMENTATION REQUIREMENTS

Before incineration process conditions can be
controlled, manually or automatically, they must be
measured with precision and reliability, as discussed
by DeMarco *et al.*[1] Instrumentation for an incineration process is essential because of the variability
of the many factors involved in attaining good combustion. For example, as the heat content of the
solid waste rises, changes in the combustion process
become necessary. Instrumentation indicates these
variations so that automatic or manual control
adjustments can be made.

The uses of instrumentation and controls include
means of process control, protection of the environment, protection of the equipment, and data collection.
A control system must have four basic elements:
(1) the standard of desired performance; (2) the
sensor (instrument) to determine actual performance;
(3) the capability to compare actual versus desired
performance (error), and (4) the control device to
effect a corrective change. Related measurements
are recorded on a single chart--the degree and
sophistication of instrumentation depends upon
plant size and economics.

INSTRUMENTATION CONTROLS
AND INDICATORS

Instrumentation systems for municipal refuse
incinerators are discussed thoroughly by Stickley.[2]
The controls and indicators follow.

Underfire Air Control

To effect complete combustion of the refuse, an adequate amount of air must be introduced underneath the stoker. In order to assure an adequate flow of underfire air, it is desirable that the actual air flow be controlled. Air flow in the underfire air duct is controlled at a preset value.

Furnace Temperature Control

To insure complete combustion of all volatile matter, the furnace outlet temperature is usually maintained in the 1700-1900°F range. This range may vary with the furnace design and configuration and is indicative of the actual fire temperature which is raised or lowered by controlling the amount of overfire air. A thermocouple provides a signal for the controller whose output positions an overfire damper by means of an electric motor.

Furnace Pressure Control

With the furnace temperature control system continually varying the total air delivered to the furnace by the forced draft fans, a draft control system is necessary to maintain the proper pressure in the furnace. Pressure is measured at the combustion chamber sidewall by a pressure transducer. The signal from the transducer provides for the necessary increase or decrease in the speed of the induced draft fan.

Cooling Control System

To prevent damage to the dust collector and the induced draft fan, hot gases coming from the furnace at 1700-1900°F must be cooled to about 650°F, generally by banks of water sprays and/or by dilution of the gas stream by outside air. The thermocouples mounted in the spray chamber outlet produce a signal activating a controller which modulates in sequence a series of motorized valves that supply water to the spray banks, thus maintaining the desired temperature. The cooling air system operates on similar principles with a signal from the controller operating the damper motor, thus regulating the amount of outside air introduced to the system.

Dust Collector Capacity
Control System

Where a multiple unit cyclone collector is used
for fly-ash control, maximum efficiency and number
of collector units in operation should be determined
by the velocities of gases through the furnace. The
speed is measured by a d-c tachometer generator;
this speed signal actuates damper motors to place
in operation the proper number of collector units
for most efficient fly-ash control.

Pressure Indication

Pressure monitors are desirable at the following
points: (a) underfire air duct, (b) stoker compartments (one for each zone), (c) overfire air duct,
(d) sidewall air duct if used (one for each side),
(e) sidewall low furnace outlet, (f) dust collector
(differential across collector), (g) induced draft
fan inlet, (h) dust collector inlet (with interlock
switch).

Temperature Indication

Similar monitorings of temperatures are desirable
so the operator may know the temperatures throughout
the facility. Such temperature monitorings should
include: (a) furnace outlet, (b) settling chamber,
(c) spray chamber outlet, (d) dust collector inlet,
(e) dust collector outlet, (f) stack.

Flow Indication

Critical air flow information should be available to the operator to keep the incinerator at peak
performance. Included here are underfire air flow
to each stoker zone, total underfire air flow, and
total overfire air flow. With more stringent
enforcement of air pollution regulations possible,
operation must be more closely controlled.

Smoke Density

In order to guard against violation of air pollution regulations, it is desirable that the operator

be continually informed as to the condition of the
gases entering the stack. Smoke density monitors
are used to continually measure the amount of smoke
in the gas stream. A light source is installed on
one side of a duct with a photocell unit on the
opposite side. Smoke and fly ash passing between
the light source and the photocell reduce the light
transmitted to the photocell, thereby reducing its
output. The output is fed to a continuously-indicating
monitor and is continuously recorded as proof of
compliance with air pollution requirements.

Alarm Systems

An annunciator panel should be installed to
indicate and identify the following malfunctions:
(a) high temperature in charging hopper, (b) high
temperature in lower drying stoker bearing, (c) low
pressure in underfire air duct, (d) low pressure in
overfire air duct, (e) high temperature at furnace
outlet, (f) high pressure at furnace outlet,
(g) high temperature at spray chamber outlet,
(h) high temperature at dust collector inlet,
(i) low water pressure, (j) high smoke density,
(k) high stack temperature.

Other Instrumentation

Since efficient operation of an incinerator is
affected by prevailing atmospheric conditions, it
may be desirable to indicate outside air temperature,
barometric pressure, wind velocity and wind direction.
Since cooling water usage may be high, it may be
economically desirable to clarify and recirculate
it. Instrumentation required for such a water treatment facility will vary with the type and extent of
treatment. Finally, proper design of any control
system must include provisions for emergency conditions such as loss of electrical power, excessive
temperatures, fan failure, etc. The necessary
interlocks must be included for protection of the
capital investment as well as for the operators'
safety.

Table 7.1 shows installed costs (1968 dollars)
per furnace unit for much of this equipment as mentioned in the work of Day and Zimmerman.[3] It is
estimated that the indicating and recording equipment

Table 7.1

Installation Costs of Incinerator Instrumentation[3]

Instrumentation	List Price	Installed Cost
Indicating Instruments on Panel		
Draft gages indicating pressures (or drafts) of all air supply systems, underfire air compartments, furnace, air pollution control equipment inlet and outlet, induced draft fan inlet and stack.	$ 800	$2,400
Indicating-Recording Instruments on Control Panel (each furnace unit)		
Smoke density recorder	1,590	2,105
Temperature recorder for furnace and stack temperature	1,975	2,900
Counter-recorder for number of buckets charged to each furnace	400	1,150
Additional Recorder for Supervisor's Panel		
Ambient air temperature	630	1,330
Wind speed and direction	2,500	4,300
Indicating-Recording-Controlling Units on Control Panel (each furnace unit)		
Furnace draft recorder controller	2,725	4,000
Flue gas temperature recorder controller	3,600	5,600
Induced draft fan motor overload controller	2,100	2,750
Remote airflow damper controller	350	700
Multiple Alarm Units (each furnace)		
Monitoring 15 alarm units for each furnace	700	1,350
Special Television Equipment		
Television cameras, six	48,000	56,000
Television monitors, three	1,670	2,270

listed could be installed, piped and wired on
appropriate panels for approximately $158,000 for
a four-furnace plant.

As discussed by DeMarco, Keller, Leckman and
Newton,[1] carefully written specifications for instrument type, quality and location, followed by
proper installation and routine testing and preventive maintenance, are keys to successful instrument
operation. Many instruments need frequent cali-
bration to ensure accurate and reliable readings.
Dust can also interfere with the working of the
instruments, and the hot and sometimes corrosive
flue gas stream can deteriorate the sensing elements
inside the gas passage. Although the instrument
responds, a testing program is necessary to verify
and maintain the accuracy of the readings.

Repair and maintenance of instrumentation often
require qualified personnel. Contract services
should be used if qualified instrument repair personnel are not available at the incinerator.
Incinerator personnel, however, should be trained
to identify and correct everyday problems, such as
clogging of transmission lines, fouling and damaging
of sensing devices, and improper charting and inking.
A maintenance and repair service contract to correct
daily problems is not warranted when the expense and
the time lag from reporting the malfunction to its
correction are considered. The incinerator personnel
should be trained in the use and interpretation of
data received from the instruments. Even if the
operators do not use the data directly, knowing the
intended use may motivate the operator to obtain an
accurate reading.

In industrial as well as municipal incineration,
an incinerator should include the instrumentation
necessary for determining the weight of incoming and
outgoing material; overfire and underfire air flow
rates; selected temperatures and pressures in the
furnace, along gas passages, in the particulate collectors, and in the stack; electrical power and
water consumption of critical units; and grate
speed. Smoke density should be continuously
monitored.[2]

Improvements in and wider application of instrumentation and controls hold promise for upgrading
routine operations and for lowering the operating
costs of incineration. Although the application of
present-day instrumentation and control technology
can improve the state of the art, new concepts, as
well as designs and applications, for controlling

the combustion process are needed. Certainly the need for improvements in measuring and controlling the weight input to the furnace is recognized. An improved device for monitoring smoke and particulate emissions is also needed. Further research is also required to understand the limitations of instruments, to ascertain the best instrument locations and to better correlate the instrument readings with incinerator performance.[1]

REFERENCES

1. DeMarco, J., D. J. Keller, J. Leckman, and J. L. Newton. "Incinerator Guidelines," Public Health Service Publication No. 2012, Office of Solid Waste Management Programs, Environmental Protection Agency (1969).
2. Stickley, J. D. "Instrumentation Systems for Municipal Refuse Incinerators," paper prepared for 1968 National Incinerator Conference (ASME) at New York (May 5-8, 1968).
3. Day and Zimmerman Associates. "Special Studies for Incinerators in Washington, D.C.," Public Health Service Publication No. 1748, Office of Solid Waste Management Programs, Environmental Protection Agency.

CHAPTER 8

CONTROL OF AIR POLLUTION

INCINERATOR EFFLUENTS

Improper design and operation of an incinerator can bring about pollution of air, water and land. Strict air and water pollution legislation at all levels of government, coupled with the trend to locate incinerators close to the sources of solid waste (to reduce hauling cost), requires an overall upgrading of the incinerator process to ensure that it does not adversely affect the environment.[1]

Odor, Dust and Litter

The operation of the tipping and storage area of municipal incinerators can cause dust, litter and noxious odors. The dust and odors generated can cause extremely unpleasant working conditions. Frequent sweeping of the tipping floor effectively controls litter. Washing floors with cleaning disinfectant solutions and frequently removing putrescibles from the pit floor aid in direct control of odors. In Europe, strategically placing combustion air intakes within an enclosed tipping area has met with some success in controlling dust and odor.

Residue from Combustion

Residue consists of all solid materials remaining after burning. It includes ash, clinker, metal cans, glass, rocks and unburned organic substances. Residue from incineration of municipal solid waste commonly may be as much as 20 to 25% by weight of the original solid waste. Uncompacted residue occupies 10 to 20% of the original volume of solid waste in the pit.

Incinerator residue is permeable and may contain water soluble organic and inorganic compounds. Because pollution can occur if there is leaching when water moves through the residue and enters the ground water, it is almost mandatory to use only sanitary landfill methods to dispose of incinerator residues. At the present time, there are no specific and universally accepted quality standards for residue from municipal incineration. Residue containing less than 5% combustibles, measured in terms of total dry weight of residue, and having a total volume of less than 10% of the original solid waste charged may, however, be acceptable from the standpoint of volume reduction in most locations. The degree of burnout will also affect the degree of protection afforded environmental quality.

Fly Ash

One of the products of combustion is fly ash. This portion of the residue consists of the solid particulate matter carried by the combustion gases. Fly ash includes ash, cinder, mineral dust and soot, plus charred paper and other partially burned materials. The size of most fly ash particles ranges from 120 to less than 5 microns. Distribution within this range is extremely variable. The inorganic fraction of fly ash is usually the major constituent and consists mostly of oxides of silicon, aluminum, calcium and iron, together with some chlorides and sulphates.

Collected fly ash may be transported in a water slurry or handled in a dry state. Fly ash process water has large amounts of solids (average 35%) and its pH is usually greater than 10. Because of these characteristics, sluicing water is treated before final disposal. Usually, it can be handled with the residue process water. At the incinerator plant, dry fly ash should be stored in suitable closed containers. If stored in the open, the surface of the ash pile should be kept moist. When transported to the final disposal site, fly ash should be in closed containers unless intermixed at the incinerator with moist residue. Sanitary landfill methods are preferred in the disposal of fly ash.[1]

A prototype fly ash monitor for municipal incinerator stacks is described by Jackson, Lieberman, Townsend and Romanek.[2] In anticipation of stricter legislative control of emissions, an instrument was

designed to separate fly ash from a measured mass of
sampled gas by a cyclone and filter combination and
to measure its mass by a beta gauge technique. Current technology makes the measurement by obscuration
meters, which are sensitive to changes in size distribution of the fly ash, and by visual observations
as the emissions leave the stack--the Ringelmann
reading. Neither of these is regarded as suitable
to effect control of the burning conditions necessary
to maintain stack effluents at or below a selected
mass loading. The new instrument comprises four
functional units: a sample probe and pitot tube, a
sample separator and mass assessment meter, a pump
station, and a remote readout and control. It is
stated to have immediate potential for simple
nonisokinetic sampling of stack effluents for
effective furnace control and, with further probe
development, may be useful in continuously monitoring
compliance with air pollution codes. It also has
high precision for measurements of dust loading
samples under automatic isokinetic and fixed
velocity conditions.

Process Water

Almost without exception, all incinerator plants
use water for residue quenching. In addition, many
plants use water for wet bottom expansion chambers,
for cooling charging chutes, for fly ash sluicing,
for residue conveying and for air pollution control.
The quantity of water required depends on plant
design, on how well the system is operated and on
whether or not water is recirculated. Because of
extreme variation in incinerator design, generalizing
on water requirements has limited value. A rule of
thumb is that residue quenching and ash conveying
at many plants requires 1,000 to 2,000 gallons of
water per ton of solid waste processed. With water
treatment and recirculation, total water consumption
can often be reduced 50 to 80%. Studies have shown
that incineration process water contains suspended
solids, inorganic materials in solution and organic
materials that can contribute to biochemical oxygen
demand. Since incinerator process waters can be
contaminated, they should not be discharged indis-
criminately to streams or other open bodies of water.
The most straightforward control is the discharge of
these waters to a sanitary sewer for subsequent
handling in a central treatment plant. If the waste
process waters cannot be ultimately discharged to a

sanitary sewer, the incinerator plant should be
equipped with suitable means for primary clarifica-
tion, pH adjustment and, if necessary, biological
treatment to meet local standards.[1]

AIR POLLUTION CONTROL

In evaluating the role of incineration as a
contributor to air pollution, it is instructive to
study the data published in 1969 by the National
Air Pollution Control Administration[3] for the principal sources and types of air pollution in the
United States for 1966. These data are given in
Table 8.1.
Of the five major categories of sources of air
pollution, refuse disposal is by far the lowest
contributor, yielding 7.9 million tons per year or
4.2% of the total. Of this, the bulk of the pollutants are due more to inadequate design and operation
of the incinerators than to inherent factors in the
disposal methods, especially as regards hydrocarbons
and particulates. It is also interesting to note
that hydrochloric acid is not one of the types of
air pollution mentioned, which suggests that the
federal agencies are aware of the relatively low
level of this material.

Particulate Matter

In fundamental terms, Pearl[4] describes the
variety of equipment used for collection of particulate matter. Briefly, these are subsidence chambers,
mechanical cyclones, wet scrubbers, filter bags and
electrostatic precipitators. From an air pollution
standpoint, particulate matter is any material that
exists as a solid at standard conditions. Some
examples of particulates are smoke, dusts, fumes,
mists and sprays. Devices for control of particulate
matter are available in a wide variety of designs
using various principles of operation and having a
wide latitude in collection efficiency, initial cost,
operating and maintenance costs, space, arrangement
and materials of construction. Devices for control
of particulate matter have been grouped by Danielson[5]
into six classes: (1) inertial separators, (2) wet
collection devices, (3) baghouses, (4) single-stage
electrical precipitators, (5) two-stage electrical
precipitators, and (6) other particulate collecting
devices.

Table 8.1

Principal Sources and Types of Air Pollution in the United States, 1966[3]

Sources	Carbon Monoxide	Sulfur Oxides (SO_x)	Types Hydrocarbons	Particulates	Nitrogen Oxides (NO_x)	Total
	Millions of tons per year					
Transportation	71.2	0.4	13.8	1.2	8.0	94.6
Fuel combustion (stationary)	1.9	22.1	0.7	6.0	6.7	37.4
Industrial process	7.8	7.2	3.5	5.9	0.2	24.6
Miscellaneous	8.6	0.6	6.5	7.2	1.4	24.3
Refuse disposal	4.5	0.1	1.4	1.2	0.7	7.9
All Sources	94.0	30.4	25.9	21.5	17.0	188.8
	% by weight of total by source					
Transportation	75.8	1.3	53.3	5.6	47.1	50.1
Fuel combustion (stationary)	2.0	72.7	2.7	27.9	39.4	19.8
Industrial processes	8.3	23.7	13.5	27.4	1.2	13.0
Miscellaneous	9.1	2.0	25.1	33.5	8.2	12.9
Refuse disposal	4.8	0.3	5.4	5.6	4.1	4.2
All Sources	100.0	100.0	100.0	100.0	100.0	100.0
	% by weight of total by type					
Transportation	75.3	0.4	14.6	1.3	8.4	100.0
Fuel combustion (stationary)	5.1	59.1	1.9	16.0	17.9	100.0
Industrial processes	31.7	29.3	14.2	24.0	0.8	100.0
Miscellaneous	35.4	2.5	26.7	29.6	5.8	100.0
Refuse disposal	56.9	1.3	17.7	15.2	8.9	100.0
All Sources	49.8	16.1	13.7	11.4	9.0	100.0

Inertial Separators

These are the most widely used devices for collecting medium and coarse-sized particulates; they
operate by imparting centrifugal force to the particle
to be removed from the gas stream. Included in the
category are: single-cyclone separators; high-
efficiency cyclone separators (work well in 5 to
10 micron range); multiple-cyclone separators (90%
efficiency for 5 to 10 micron range); and mechanical
centrifugal separators in which the centrifugal force
is supplied by a rotating vane. Efficiencies of the
mechanical centrifugal types are higher than those
with simple cyclones. They can be of great value if
used in conjunction with other equipment such as an
electrostatic precipitator to collect the smaller
particles.

Wet Collection Devices

Wet collection devices use a variety of methods
to remove solid particles from the gas stream, including impingement by spray droplets, diffusion and
condensation. They vary in cost over a wide range,
in collection efficiency and in the amount of power
needed, and include spray chambers, cyclone-type
scrubbers, orifice-type scrubbers, mechanical scrubbers, mechanical and centrifugal collectors with wet
sprays, high pressure sprays, Venturi scrubbers, packed
towers and wet filters. In general, the collection
efficiency is proportional to the energy required, and
since high energy devices are expensive to install and
operate, most incinerators in use to date have limited
efficiency. [5]

Baghouses

When high collection efficiency of small particles
is required, suspended dust and fumes respond well to
this method, which consists of separating the dust
from the air by means of a fabric filter. The fabric
is usually made into bags of tubular or envelope
shape--the entire structure housing the bags is called
a baghouse. In operation, small particles are initially
captured and retained on the fibers of the cloth by
means of interception, impingement, diffusion, gravi-
tational settling and electrostatic attraction. Fibers
used as filtering media include cotton, dynel, wool,
nylon, orlon, dacron and glass, all with variations
in weave, count, finish, etc. Other variations include the size and shape of filters, arrangement and

spacing of bags, and method of attachment to the
support. Based on experience with an installation of
a pilot baghouse on the municipal incinerator of the
city of Pasadena, California, the advantages are:
(1) high efficiency--99%; (2) moderate press drop,
3 in. to 5 in. water; (3) the filtering out of both
small and large particulates; (4) the ability to
filter out SO_3 due to the nature of ash cake on the
bag filter. The disadvantages are: (1) high initial
cost, (2) costly bag replacement, (3) requirement of
greater control of combustion to eliminate sticky
soot formation which clogs filters, (4) necessity to
control cooling to prevent formation of moisture on
the filter which will shorten bag life. Baghouses
for municipal incinerators are not in general use.

Single-Stage Electrical Precipitators

Electrical precipitation is frequently called
the Cottrell process and is defined as the use of
an electrostatic field for precipitating or removing
solid or liquid particles from a gas in which the
particles are carried in suspension. The equipment
used for this process is called a precipitator or
treater in the United States and an electrofilter
in Europe. After Cottrell proved that electrical
precipitation could be applied successfully to the
collection of industrial air contaminants, the use
of electrical precipitation expanded into many
diverse fields. The process of electrostatic precipitation consists of a number of elements or
mechanisms: (1) gas ions are formed by means of
high voltage corona discharge; (2) the solid or
liquid particles are charged by bombardment by the
gaseous ions or electrons; (3) the electrostatic
field causes the charged particles to migrate to a
collecting electrode of opposite polarity; (4) the
charge of a particle must be neutralized by the
collecting electrodes; (5) reentrainment of the
collected particles must be prevented; (6) the
collected particles must be transferred from the
collecting electrode to storage for subsequent
disposal. The accomplishment of these functions
by an electrical precipitator has required the
development of many specialized techniques for
specific materials and is suitable for the collection of a wide range of dusts, fumes and dispersoids.
In general, the size of particulate matter varies
from 5 to 100 microns, and efficiencies considerably
in excess of 99% can be achieved.[5]

Two-Stage Electrical Precipitators

The Cottrell-type precipitator is usually designed
and built for installations required to process large
volumes of contaminated air. The two-stage precipitator is sometimes called the low-voltage, the Penney,
or the electronic air filter. It differs from the
Cottrell type in that the contaminated air is passed
through a variable strength ionizing field before
being subjected to a separate uniform field where the
particles are collected. Basic operating principles
are the same as those for the single-stage precipitator. To date the experience in the United States
with electrostatic precipitation on municipal incinerators is considerably less than in Europe, the
tendency being to avoid the high capital costs by
using simpler devices such as wet scrubbers and water
sprays. As the need to reduce still further the amount
of particulate emissions increases, it is believed that
electrostatic precipitation will receive greater
attention.

Other Particulate Collecting Devices

This classification includes settling chambers,
impingement separators and panel filters. These are
used more as precleaners, but they do find uses as
final collectors where the air contaminant is large.[5]

Control Equipment for Gases and Vapors

Unburned hydrocarbons and carbon monoxide from
the main combustion chamber of the incinerator as
well as the inorganic acidic gases such as hydrogen
chloride, sulphur oxides and nitrogen oxides, which
arise from the incineration process, should be minimal
in any properly designed and operated equipment, as
should other organic vapors such as organic acids,
but it must be admitted in practice that such
materials are often to be found in the flue gases.[5]
As discussed,[5] control equipment useful here
includes: (a) afterburners--direct-fired and catalytic, (b) adsorption apparatus, (c) vapor condensers,
(d) gas absorption apparatus.

Afterburners. The direct afterburner is used
as an air pollution control device in a large variety
of industrial and commercial equipment whenever such
equipment releases combustible aerosols, gases or
vapors into the atmosphere. The catalytic afterburner finds its place in the control of organic
vapor emissions, including solvents, exhausted from
industrial ovens and other operations.

Adsorption Equipment. By adsorption, gases, liquids or solids, even at very low concentrations, can be selectively captured or removed from air streams with specific materials known as adsorbents. The material adsorbed is called the adsorbate. Among the most used adsorbents are activated carbon, silica gel, alumina and bauxite. Activated carbon is the material most suitable for removing organic vapor.

Vapor Condensers. Air contaminants can be discharged into the atmosphere in the form of gases or vapors. These gases or vapors can be controlled by several methods--in many instances, control can be accomplished by condensation. Condensers may be either surface or contact in type. In the former, the coolant does not contact the vapors or condensate, whereas in contact types, coolant, vapors and condensate are intimately mixed. In comparison with surface condensers, contact types are more flexible, simpler, and considerably less expensive to install. On the other hand, the surface type requires far less water and produces 10 to 20 times less condensate. Contact condensers normally afford a greater degree of air pollution control because of condensate dilution. With direct-contact units, about 15 pounds of 60°F water is required to condense 1 pound of steam at 212°F and to cool the condensate to 140°F.

Gas Absorption Equipment. From an air pollution standpoint, absorption is useful as a method of reducing or eliminating the discharge of air contaminants to the atmosphere and involves a mechanism whereby one or more constituents are removed from a gas stream by dissolving them in a selective liquid solvent. Some gaseous air contaminants controllable by absorption include sulfur dioxide, hydrogen sulfide, hydrogen chloride, chlorine, ammonia, oxides of nitrogen, and light hydrocarbons. Absorbers that disperse liquid include packed towers, spray towers or spray chambers and Venturi absorbers. Equipment that uses gas dispersion includes tray towers and vessels with sparging equipment. Included in the category of tray towers is the bubble cap plate tower design.

Evaluation of Air Pollution Control
Equipment in Municipal Incineration Systems

Day and Zimmerman Associates[6] give an evaluation of several control systems, summarized as follows.

Settling Chamber

Here, the velocity of the flue gases is reduced permitting the larger particles to settle out. Gas velocities in a settling chamber must be extremely low if particles smaller than 30 microns are to be separated.

> 30-micron particle -- settling velocity = 10 ft/min
> 10-micron particle -- settling velocity = 1 ft/min
> 1-micron particle -- settling velocity = 1/4 ft/min

Baffles may be inserted upon which the particles will impinge; retention is improved by wetting the baffles, with ash removal accomplished by flushing the ash into a wet sump.

The advantages are: simplest method of fly ash control, low maintenance cost, and capability of being operated with natural-draft chimney. Disadvantages are: large size, high installation cost, low collection efficiencies of 40 to 60%, and unsuitability for collection of smaller particles. Due to low efficiency, settling chambers alone will not meet the air pollution code requirements.

Mechanical Cyclones

Here, the particles are thrown to the periphery of the cyclones by centrifugal force and allowed to settle out. Advantages are low initial cost and low operating cost. Disadvantages are low efficiency (only larger particles are efficiently removed), erosion of the lower tube by abrasive fly ash, and problems with moisture control.

Flushing of the fly ash with water has been tried in the hydrowall cyclone, a developmental modification of the conventional cyclone which is reported to prevent reentrainment of particles, reduce erosion and improve efficiency. While mechanical cyclones do not completely meet the air pollution code requirements, they can be of great value when used in conjunction with an electrostatic precipitator.

Wet Scrubbers

There are two general classifications of wet scrubbers, the low energy type and the high energy type.

In the low energy scrubber, water is coarsely sprayed over the gas stream removing the particulates from the gas phase. Low energy scrubbers have low maintenance and low cost, both initial and operating,

but also low efficiency. With water droplets larger
than 200 times the diameter of the particles, the
particles will not be effectively removed from the
effluent stream, and efficiency for removal of water
soluble gases is low because of the limited amount
of contact of the gas stream with the scrubbing
liquid.

   In a high energy scrubber the water sprays are
fine and are distributed more evenly. The gas stream
path is more tortuous because of the insertion of
baffles by the use of, for example, packing.

   The advantages of a high energy scrubber include:
(1) efficiency of 95% or greater, (2) removal of
water soluble gases, (3) moderate cost of installation
and operation. The disadvantages are: (1) high
maintenance cost, (2) potential corrosion, (3) need
for equipment to remove particulate matter from
washwater and to neutralize the water before return
to source, (4) difficulty with water recirculation
because of particulates, (5) requirement for large
flow of water, (6) a water-saturated gas is emitted
from the stack, resulting in a large vapor plume
which can be objectionable. Where the vapor plume
is not objectionable, high energy scrubbers can
meet the requirements of pollution control equipment
for municipal incinerators.

   Two methods can be considered for elimination of
the vapor plume. The first method consists of re-
heating the stack gases to increase the temperature,
permitting a greater dispersion of the gases in the
atmosphere before condensation takes place. The
second method consists of cooling the gases to
approximately 100°F which will reduce the total
moisture content of the gas stream to about one-tenth
of that present in the saturated gases at 170°F.
This reduces the plume's total water content.

   Water conditioning equipment must be provided
to process the large quantities of water required
for a high energy wet scrubber. One equipment
manufacturer requires approximately 750 gallons of
water per minute (gpm) for a scrubber and quench
unit designed to handle the products of combustion
from a single incinerator furnace of 400 ton/day
size. Assuming that this water is pumped from a
nearby river, a river water intake of 3,000 gpm
capacity, suitable filtration equipment to remove
solids which might plug spray nozzles and water
treatment equipment to remove the fly ash from the
scrubber discharge water and to neutralize the acid
content of the water would be required. Total water

requirements to include additional cooling of the
gas to reduce the plume are estimated at 6,000 gpm.
The temperature rise of this would be be 100°F.
This would double the size of the intake structure,
pumping equipment, clarifier equipment, pipelines,
etc. This approach is not justified if the plume
can be accepted. Likewise, gas-to-gas heat exchangers to raise the temperature of the gases are
quite expensive to maintain and operate and their
use is not recommended.[6]

Electrostatic Precipitators

In the operation of electrostatic precipitators,
the particles are first electrically charged and
then attracted to plates which have an opposite
charge. The particles lose their charge upon contact with the plates and migrate down to collection
hoppers. The efficiency is dependent on the resultant vector between the inertia of the particles and
the electrostatic attraction to the plates.

The advantages of electrostatic precipitators
are as follows: (1) low operating cost, (2) high
efficiency (90 to 95%), (3) higher efficiency for
particles smaller than 10 microns, (4) ability to
handle both dusts and mists.

The disadvantages are: (1) high purchase and
installation costs; (2) necessity of uniform gas
distribution across the inlet of the collector to
obtain design efficiency; (3) critical electrode
voltage (too little reduces efficiency and too much
causes electric arcing); (4) two limiting factors
related to velocity and therefore capacity--particles
must have time to build up charge and gas velocity
must be low enough so as not to reentrain particles;
(5) the tendency of carbon to lose its charge before
it is collected and the difficulty in charging highly
resistant inorganics (This can be corrected by the
insertion of a cyclone before the precipitator which
will remove particles greater than 10 microns and
the addition of moisture to reduce the resistance of
the inorganics); (6) critical temperature (optimum
temperature range is 500° to 600°F because of
resistance of particles to being charged at higher
or lower temperatures).

Electrostatic precipitators should meet the requirements of pollution control equipment for
municipal incinerators. In the application of
electrostatic precipitators to municipal incinerators, serious consideration must be given to

potential operating problems. These problems could consist principally of erosion, corrosion and fouling, and passing of large particulate matter. It is probable that the problems of erosion, fouling by fatty acid particulates, and passing of large particulates can be reduced to acceptable levels if the electrostatic precipitator is preceded by a large-diameter mechanical cyclone collector constructed with an abrasion-resistant lining. This will result in a total draft loss equivalent to the wet scrubber installation.

The problem of corrosion can be reduced by good temperature control equipment and adequate insulation of the equipment to reduce internal dew point condensation. There will remain a possibility of some fouling due to accumulations of cementitious fly ash. This must be considered a normal operating problem and the equipment will require internal cleaning at scheduled intervals along with other routine maintenance such as replacement of electrode wires.

The use of alkali cleaners would be effective in removing any fatty acid films. Alkaline solutions can be used satisfactorily on carbon steel and stainless alloy steels. Wash solutions are normally used between 140° to 200°F and can be used as a spray. Wash should be followed with a rinse water spray.[5]

Baghouse Filters

In the operation of baghouse filters, the gases pass through the bag filter and the large particles are filtered out. After a few seconds, the large particle buildup on the bag enables the smaller particles to be filtered out.

Personnel connected with the installation of a pilot baghouse on the municipal incinerator of the city of Pasadena, California summarized their comments as follows: (1) The flow through the baghouse is opposite to that of conventional units. The gas enters the bag on the outside and exits through the partially collapsed bag upward through the plenum. This scheme permits the spider framework to be on the clean side of the bag and eliminates past failures due to abrasion from buildup on the spider. It also has the advantage of eliminating bag damage due to collapsing the bag over hard cakes that are periodically formed

within. Since the bag is normally in a semirelaxed
position, the cakes that are formed can be literally
"popped off" with no damage to the bag when it is
inflated (cleaning cycle). (2) The flue gas should
not exceed 500°F so as to extend bag life. If this
is accomplished bag life should be about 1 year.
(3) The unit cycle cannot go below dew point because
of bag life. (4) Pulsating damper speed in the
cleaning fan discharge is very critical (a minimum
of 250 rpm to a maximum of 500 rpm has some merit.
(5) Internal framework should be fabricated from
316 stainless steel. Carbon steel oxidizes and
results in bag failure. Aluminum does not withstand
prolonged elevated temperatures. The use of a baghouse filter for air pollution control equipment
for municipal incinerators should not be considered
at this time due to lack of sufficient satisfactory
full-scale experience.[5]
  Other material on qualities of air pollution
control equipment can be found in DeMarco *et al.*,[1]
Fernandes,[7] Jones and Henry Ltd. Engineers,[8] and
Fife and Boyer.[9]

Venturi Scrubbers

  In considering a wet-approach Venturi scrubber,
Ellison[10] describes in detail this closed-loop system
installed on the 73rd Street Manhattan (New York City)
incinerator. Earlier experience with wet-type flue
gas cleaning systems, similar to the Venturi scrubbing
system put into service by New York City's Environmental Protection Administration on June 25, 1969,
clearly indicated that soluble acid gases like
hydrogen chloride may be efficiently absorbed in
such closed-loop gas scrubbers. In addition,
important nonsoluble acid gases in incinerator flue
gas, including sulfur dioxide, may also be removed
by adding inexpensive alkali-chemical to the
scrubbing liquid. Moreover, experience with earlier
scrubber systems at municipal incinerators in
several other eastern cities has shown that almost
all of this alkali-chemical requirement may be met
by washing of waste alkali contained in the ash
residue discharged from grates of the incinerator
furnace. The special efforts of the City of New
York in the late 1960's in pioneering the use of
the wet-approach Venturi scrubber for cleaning of
municipal incinerator flue gas offer an advanced
standard of performance to the incinerator industry

for the control of air pollution. This technological
advancement is of particular importance in that it
provides a reliable and efficient means of capturing
the corrosive acid gases liberated in the disposal-
by-incineration of municipal refuse.[10]

Relative Costs of Air Pollution
Control Equipment for Incinerators

Fernandes[7] states that it is extremely difficult
to precisely pinpoint the prices of particular classes
of air pollution control equipment, since they vary
substantially from one vendor to another and with
market conditions. Furthermore, designed-in improvements cost more than less sophisticated designs of
the same class of equipment. All values must be
considered as estimates representative of a range of
possible values.
The approximate cost of the control equipment
per ton-day of rated incinerator capacity can be
developed assuming an inlet gas temperature of
600°F, 150% excess air, and the heat value of refuse
at 5000 BTU/lb. The volume of gas handled is
approximately 520 cubic feet per minute (cfm) per
ton per day of capacity if the gas cooling is
accomplished with water quenching to 600°F. Cooling
by other means (*e.g.*, by heat exchangers to generate
steam, hot water or heated air) drops the gas volume
to 365 cfm. Further reduction in gas volume to 220
cfm would occur were the combustion tube completed
with 50% excess air and a heat exchanger used to
cool the gas to 600°F.
The most authoritative data found came from
statements given before the U.S. Senate Committee
on Public Works, Subcommittee on Air and Water
Pollution, May 18, 1967, by Mr. Earl L. Wilson,
President of the Industrial Gas Cleaning Institute,
Inc. These data are shown in Table 8.2--all other
reliable information falls within these broad
ranges (figures are in 1967 dollars).
It should be noted that because of the difficulties involved in the collection of incinerator fly
ash, the actual cost of incinerator air pollution
control equipment would probably be found at the
higher levels of the ranges given. These cost
ranges also account for the decrease in price with
increase in unit size.
In illustrating how the data are used, a 95%
efficient electrostatic precipitator on a

Table 8.2

Cost Data for Equipment[7]

	Equipment Cost ($/cfm)[a]	Erection Cost ($/cfm)	Yearly Maintenance & Repair Cost ($/cfm)
Mechanical collector	$0.07-$0.25	$0.03-$0.12	$0.005-$0.02
Electrostatic precipitator	0.25- 1.00	0.12- 0.50	0.01 - 0.025
Fabric filter	0.35- 1.25	0.25- 0.50	0.02 - 0.08
Wet scrubber	0.10- 0.40	0.04- 0.16	0.02 - 0.05

[a]Dollars per actual cubic foot of gas handled.

400-ton-per-day incinerator operated in accordance with the foregoing assumptions would have the following cost:

    520 cfm/ton per day x 400 tons/day = 208,000 cfm
    208,000 cfm x $0.80 = $166,400 F.O.B. manufacturing
                                    plant
    208,000 cfm x $0.40 = $ 83,200 delivery and cost of
                                    erection
                          $249,600 total delivered and
                                    erected cost
    208,000 cfm x $0.02 = $ 4,160 yearly maintenance and
                                    repair cost

    A summary of the comparative air pollution control data for municipal incinerators is shown in Table 8.3. In this table, the second column gives the space needed for each class of system. Column six gives the very important comparison of the relative operating cost among the various systems. Many communities buy their units on a lowest capital cost basis without regard for the continuing operating expense. On a capital cost basis, it would be difficult to justify a unit with improvements such as indirect heat exchange. Units of this type, when energy credits are not included, are at best only on a cost par with water quench systems.
    Day and Zimmerman Associates[6] discuss two concepts for the removal of air pollutants from

Table 8.3

Comparative Air Pollution Control Data for Municipal Incinerator[7]

Collector	1 Relative Capital Cost Factor (F.O.B.)	2 Relative Space (%)	3 Collection Efficiency (%)	4 Water to Collector (per 1000 cfm)	5 Water Column Pressure Drop (In.)	6 Relative Operating Cost Factor
Settling Chamber	Not Applicable	60	0.30	2.3 gpm	0.5.1	0.25
Multicyclone	1	20	30.80	None	3.4	1.0
Cyclones to 60" Dia. Tangential Inlet	1.5	30	30.70	None	1.2	0.5
Scrubber[a]	3	30	80.96	4.8 gpm	6.8	2.5
Electrostatic Precipitator	6	100	90.97	None	0.5.1	0.75
Fabric Filter	6	100	97-99.9	None	5.7	2.5

[a]Includes necessary water treatment equipment.

incinerator flue gases which were made the subject
of capital investment and operating cost estimates.
In both cases, the gas flow is created by an induced
draft fan which discharges to a 100-foot chimney.
The efficiencies and basic limitations of electrical,
mechanical and wet scrubber equipment, which might
be used in such systems, are summarized in Table 8.4.

Table 8.4

Types of Control Equipment[6]

Equipment Type	Comparative Space (%)	Efficiency (%)	Basic Limitations
Electrostatic precipitator	100	90-99	Does not remove soluble gases. Efficiency low on large particles.
Scrubber (flooded plate)	33	90-99	Possible mist emitting from stack. Clarification and neutralization of wash water required. High water usage.
Mechanical cyclone (60 in. tangential)	33	75-90	Low efficiency on small particles, erosion from abrasive fly ash.
Baghouse filter	110	99	Multitude of variables to be controlled coupled with the complexity of the effluent stream.
Setting chamber	67	40-60	Low efficiency.

One arrangement provides for passing the 1,260°F
flue gas from the furnace through a refractory
spray-cooling chamber where the temperature is reduced to 500°F, then through a multicyclone separator,
and finally at a low velocity of 5 feet per second,
through an electrostatic precipitator unit.
The alternate concept consists of passing the
hot furnace gas directly through a wet scrubber such
as the flooded-plate type with a prequench unit.
Water discharge from the scrubber would be delivered
to a liquid clarifier. Overflow would be conducted
back to the water source by pipeline (in this case
river water is considered). Slurry removed would be
pumped to residue trucks for landfill disposal.

Either of these systems is capable of removing
94% or more of the particulate matter of a size
greater than 10 microns, and close to 100% of
particulates less than 10 microns in size.

Annual operating costs and estimates of owning
costs are shown in Table 8.5.[6] Building costs,
the spray cooling chamber, steel ductwork, insulation, electrical work, and instrumentation add
appreciably to the cost of installation. Only those
variables are included which are influenced by the
type of equipment considered.

Table 8.5

*Estimated Annual Operating Costs for Two Types
of Air Pollution Control*[6]

| Operating Costs | Air Pollution Control Equipment | |
	Electrostatic and Mechanical	Wet Scrubber
Maintenance	$139,000	$127,700
Electric power	160,600	130,800
Purchased city water	27,900	1,000
Subtotal, operating costs	327,500	259,500
Fixed charges on capital investment 20 years at 4 1/2%	185,000	141,500
Total, annual owning and operating costs	$512,500	$401,000

The capital costs for the type of air pollution
control equipment under consideration in the study
are shown in Table 8.6.

Fife and Boyer[9] present an objective but statistical approach on a number of possible combinations
of air pollution control equipment for municipal
incinerators. Both refractory-lined and water-walled
furnaces of identical capacities are followed by gas
tempering systems, where required, and thence by
mechanical cyclones, electrostatic precipitators,
bag filters, or alternate equipment such as refractory-
lined baffled spray chamber or by a wet scrubbing

Table 8.6

Comparative Capital Cost Estimates of Selected Items
for Air Pollution Control Study, Four-Unit Incinerator Plant[6]

| | *Air Pollution Control Equipment* | |
Item	*Electrostatic and Mechanical*	*Wet Scrubber*
General building contract: Incinerator building and foundations	$ 606,700	$ 447,000
River pump house		7,200
Clarifier basins		45,000
	$ 606,700	$ 499,200
Mechanical contract: Refractory furnaces and flues	204,000	248,400
Spray cooling chamber	221,900	
Steelwork	324,000	228,000
Insulation	133,700	26,500
Instrumentation	60,000	42,000
Installation of purchased equipment	132,000	60,000
Piping		58,500
	$1,075,600	$ 663,400
Purchased equipment: Fans and drives	135,900	125,900
Pumps and drives		19,000
Air pollution control equipment	396,000	337,300
Clarifier equipment		64,800
	$ 531,900	$ 547,000
Electrical contract: Power and Lighting	$ 195,000	$ 129,000
Subtotal physical cost	$2,409,200	$1,838,600
Eng. and field supervision	169,000	130,000
Contingency	241,000	185,900
Escalation to December 1968	120,500	92,900
Total incremental physical cost	$2,939,700	$2,247,400

system, with the furnace gases going directly to the cleaning equipment in these latter two cases. Table 8.7 shows the data on eight of the systems studied.

Table 8.7

Economic Appraisal of the Costs of Constructing, Owning, and Operating Air Pollution Equipment to Meet Various Municipal Stack Emissions[9]

System	Average Construction Cost (1966 dollars)	Dollars/ton of Refuse Burned	Stack Outlet lb/1000 lb Flue Gas @ 50% Excess Air
(A) Baffled spray chamber	188,200	0.77	1.75
(B) Spray chamber/ cyclone collector	270,360	1.23	0.77
(C) Wet scrubber	400,900	2.10	0.14
(D) Spray chamber/ electrostatic precipitator	501,770	1.21	0.175
(E) Spray chamber/ bag filter	712,190	2.00	0.035
(F) Water cooled furnace/cyclone collector	91,800	0.38	0.77
(G) Water cooled furnace/electro- static precipitator	210,300	0.39	0.175
(H) Water cooled furnace/bag filter	243,000	0.65	0.035

Another reference of interest is Walker and Fritz[11] in which "optical pollution" is discussed with excellent coverage given to mechanical collectors, scrubbers (most of the known types are discussed), fabric filters, electrostatic precipitators, adsorption equipment, absorption equipment, (rotating fixed-bed absorber, packed-tower absorber), combustion (flame and catalytic) and chimneys, all with both technical and cost considerations. Niessen[12] gives tabular and other data on annual costs, total installed costs and cost distribution, for the types of control equipment in normal present use (or in expected future use) on municipal incinerators.

Dry Filters

The National Air Pollution Control Association
(NAPCA) contracted with Avco (under contract No.
PH-86-67 51, Phase II AVATD-0107-69-RR) for a study
of granular bed devices in the simultaneous removal
of fly ash and SO_2 gas from coal fired power
stations.[13] Considerable attention has been focused
in recent years on the removal of SO_2 from stack
gases through the development of dry processes which
would operate on hot flue gases. SO_2 removal can
be achieved by sorption of the sulphur species on
pellets of several types displaying high specific
rates and capacities, and a material such as
akalized alumina would also act as the granular
collecting particles for fly ash.

Broadly, a granular gas filter, as considered
for the purpose of this discussion, is a stationary
fixed bed or close-packed moving bed device consisting of separate, relatively close-packed granules
which make up the collection. To prevent plugging
of interstices between the granules which could cause
excessive pressure drops, the device should embody
some means for either periodic or continuous removal
of fly ash from the collecting surfaces. Although
a number of different sorbents can be considered for
SO_2 removal, the most extensively studied material
is alkalized alumina.

Some specific types covered in the Avco study
included the following designs: the Dorfan Impingo
filter, invented by M. I. Dorfan of Pittsburgh,
Pennsylvania, and once commercially sold; the
Consolidation Coal Company crossflow filter; the
Carnegie-Mellon University crossflow shaft filter
under study there; the Squires LS filter, invented
by Prof. A. M. Squires of New York (the similar
Squires GSC filter has not been experimentally tested
but holds promise as a countercurrent sorber); the
Fuller Company "Sand Bed" filter, also under development (1969); the Lurgi Company (German firm) "Gravel
Bed" filter, once marketed in Europe and undergoing
redesign as of late 1968; and the Granger Filter
Company filter. Of these, the first three are considered to be crossflow shaft-falling solid designs.
The Squires and Granger devices are considered to be
intermittent moving bed filters. The Dorfan, the
Fuller and the Lurgi (gravel-bed) filters are
considered to be fixed-bed.

Values for both capital and operating costs for
the granular devices were of the same order as
published estimates for other SO_2 removal systems.

For particulate removal, estimated costs were of the same order as the costs for electrostatic precipitation equipment.

The study indicates that insufficient data exist to permit the development of a general correlation that could be used for predicting the performance of any given proposed device. Although, under certain conditions, high efficiencies were realized, this limited work is judged insufficient to constitute a definite technical evaluation of the devices as fly ash collectors.[12]

A review of the voluminous material presented in the Avco study is given in Reference 14. Finally, Squires and Pfeffer[15] discuss panel bed filters for the simultaneous removal of fly ash and SO_2. Exploratory trials of three panel bed filters--using 16 to 30 mesh sand and operating at a face velocity around 12 ft/min--indicated a good probability that such a device can achieve 99% filtration efficiency on power-station fly ash. A "puff-back" technique was used to remove fly ash and a portion of the sand from the gas-entry face of a panel. Based upon the experimental results, a commercial design would appear to provide a gas-treating capacity per unit ground area on the order of four times greater than an electrostatic precipitator of comparable efficiency. The panel bed device also has the power to remove SO_2 if a suitable reactive filter solid is provided.

REFERENCES

1. DeMarco, J., D. J. Keller, J. Leckman, and J. L. Newton. "Incinerator Guidelines," Public Health Service Publication No. 2012, Office of Solid Wastes Management Program, Environmental Protection Agency (1969).

2. Jackson, M. R., L. Lieberman, L. B. Townsend, and W. Romanek. "Prototype Fly Ash Monitor for Municipal Incinerator Stacks," paper prepared for 1970 National Incinerator Conference (ASME) at Cincinnati, Ohio (May 17-20, 1970).

3. "Summary of Nationwide Emissions," Office of Solid Wastes Management Program, Environmental Protection Agency, Public Health Service National Air Pollution Control Administration, Cincinnati, Ohio (1969).

4. Pearl, D. R. "A Review of Modern Municipal Incinerator Equipment," Vol. IV, *Technical Economic Overview*, Part 4 *Technical Economic Study of Solid Waste Disposal Needs and Practices by Combustion Engineering, Inc.*, Public Health Service Publication No. 1886, Office of Solid Wastes Management Program, Environmental Protection Agency (1969).

5. Danielson, J. A. "Air Pollution Engineering Manual,"
    Public Health Service Publication No. 999-AP-40, National
    Center for Air Pollution Control, Cincinnati, Ohio, for
    Air Pollution Control District, Los Angeles County,
    California (1969).
6. Day and Zimmerman Associates. "Special Studies for
    Incinerators in Washington, D.C.," Public Health Service
    Publication No. 1748, Office of Solid Wastes Management
    Program, Environmental Protection Agency.
7. Fernandes, J. H. "Incinerator Air Pollution Control,"
    paper prepared for 1968 National Incinerator Conference
    (ASME) at New York (May 5-8, 1968).
8. Jones and Henry Engineers Ltd. "Proposal for a Refuse
    Disposal System in Oakland County, Michigan," Public
    Health Service Publication No. 1960, Office of Solid
    Wastes Management Program, Environmental Protection
    Agency (1970).
9. Fife, J. A. and R. H. Boyer, Jr. "What Price Incinera-
    tion Air Pollution Control," paper prepared for 1966
    National Incinerator Conference (ASME) at New York
    (May 1-4, 1966).
10. Ellison, W. "Control of Air and Water Pollution from
     Municipal Incinerators with Wet-Approach Venturi
     Scrubber," paper prepared for 1970 National Incinerator
     Conference (ASME) at Cincinnati, Ohio (May 17-20, 1970).
11. Walker, A. B. and N. W. Frisch. "Scrubbing Air,"
     Science and Technology (May–June, 1970), pp. 18, 24.
12. Niessen, W. R. *et al.* "Systems Study of Air Pollution
     from Municipal Incinerators," National Air Pollution
     Control Administration, Environmental Protection Agency,
     Contract CPA-22-69-23, Vol. 1, PB 192 378 (March, 1970).
13. Avco Corporation. "Evaluation of Granular Bed Devices,"
     Contract PH-86-67-51, Phase II AVATD-0107-69-RR. Prepared
     for Process Engineering Control Program, National Air
     Pollution Control Administration, Environmental Protection
     Agency (June, 1969).
14. "Simultaneous SO_2 and Fly Ash Removal," *Environmental
     Science and Technology 4(1)* (January, 1971), pp. 18, 19.
15. Squires, A. M. and R. Pfeffer. "Panel Bed Filters for
     Simultaneous Removal of Fly Ash and Sulfur Dioxide,"
     Journal of the Air Pollution Control Association 20(8)
     (August, 1970), pp. 534-538.

CHAPTER 9

EMISSION LEVELS FOR AIR POLLUTANTS

GENERAL

Fernandes[1] points out that the primary air pollution concern is with particulate emission rather than
gases or odors. While this may be the major concern
with municipal incineration in general, plastics may
have to be considered with other emissions, such as
acidic gases or acid precursor materials in both
municipal and industrial or on-site incineration.
Fernandes states that there have been comments to
the effect that a properly operated incinerator does
not need particulate collection equipment. Many
systems with little or no pollution control equipment
have been represented as effectively meeting dust
emission requirements when they actually do not.
This occurs because the excess air used for combustion
and cooling is so great (200 to 500%) that it dilutes
the effluent to the extent that it does not appear
objectionable, although excessive quantities of dust
are actually emitted. With the trend toward larger,
more efficient incinerators (*e.g.*, regional incineration) located close to the population centers
served, effective control of incinerator atmospheric
pollution is extremely important.

PARTICULATE EMISSIONS

Niessen and Serofim[2] point out that the amount
of particulate emissions from an incinerator are
primarily affected by two factors: (a) the composition of the refuse, and (b) the design and operation
of the incinerator.

The mechanisms mainly responsible for the particulates are: (a) the mechanical entrainment of
particles from the burning bed, (b) the cracking of

129

pyrolysis gases, and (c) the volatilization of inorganic salts or oxides.

The first of these mechanisms is favored by a refuse in which there is a large percentage of ash, particularly if this is of fine particle size, by high underfire gas velocities or other factors inducing high gas velocity through the burning refuse bed (such as nonuniform density, areas of high burn-through and so forth). The second mechanism is favored by refuse with a high volatile content producing pyrolysis gases having high carbon content and by conditions above the fuel bed preventing complete burnout of the carbon formed by the cracking of the volatiles. It will be clear that plastics may play a particularly important role in this mechanism. The third mechanism is favored by the presence of high vapor pressure metal oxides coming from the refuse constituents and by high temperatures in the incinerator.[2]

While it would seem reasonable and simple to express particulate emission in some standard way such as pounds per ton of refuse incinerated, or as grams per cubic meter (or per cubic foot) of gas emitted, the literature shows other means of expression, as is common with other emissions, such as hydrogen chloride or sulphur oxides. The additional bases of reporting are:

(a) Pounds per 1000 pounds flue gas corrected to 50% excess air.
(b) Pounds per 1000 pounds flue gas, corrected to 12% CO_2.
(c) Grams per standard cubic foot (60°F, 1 atm) at 50% excess air.
(d) Grams per standard cubic foot (60°F, 1 atm) at 12% CO_2.
(e) Grams per cubic meter at NTP (32°F, 1 atm) and 7% CO_2.

To compare data from different sources, conversion factors are needed, but since these depend on the composition of the refuse and the operating conditions in the specific incinerator, Niessen and Sarofim[2] have made certain assumptions and derived the conversion factors as follows:

(a) Refuse Composition -- see Table 9.1
(b) Conversion Factors -- see Table 9.2

Analysis of the data for some 50 incinerators having a capacity of greater than 50 tons per day

Table 9.1

Refuse Composition[2]

Category[a]	Wt % "As Fired"	Wt % "As Discarded"	% Moisture "As Discarded"	Wt	C	H₂	O₂	S	N₂	Ash
							In pounds[a]			
Metal	8.7	8.2	2.0	11.19	0.50	0.067	0.481	0.0011	0.0056	10.13
Paper	44.2	35.6	8.0	45.59	20.70	2.781	19.193	0.0547	0.1368	2.74
Plastics	1.2	1.1	2.0	1.50	0.90	0.125	0.285	0.0045	0.0150	0.17
Leather & rubber	1.7	1.5	2.0	2.05	1.23	0.170	0.390	0.0062	0.0205	0.24
Textiles	2.3	1.9	10.0	2.38	1.10	0.152	0.995	0.0048	0.0523	0.08
Wood	2.5	2.5	15.0	2.96	1.43	0.178	1.260	0.0033	0.0089	0.09
Food waste	16.6	23.7	70.0	9.90	4.13	0.574	2.730	0.0248	0.2772	2.17
Yard waste	12.6	15.5	50.0	10.79	5.31	0.701	3.890	0.0378	0.3129	0.54
Glass	8.5	8.3	2.0	11.32	0.06	0.008	0.041	-	0.0034	11.21
Miscellaneous	1.7	1.7	2.0	2.32	0.30	0.046	0.278	-	0.0696	1.62
	100.0	100.0		100.00	35.66	4.802	29.543	0.1372	0.9022	28.99

[a]Basis: 100 lb dry solids.

Table 9.2

Conversion Factors[2]

	lb/ton Refuse (As Received)	lb/1000 lb Flue Gas at 50% Excess Air	lb/1000 lb Flue Gas at 12% CO_2	gr/st ft³ at 50% Excess Air	gr/st ft³ at 12% CO_2	G/Nm³ at ntp, 7% CO_2
lb/ton refuse (as received)	1	0.089	0.10	0.047	0.053	0.067
lb/1000 lb flue gas at 50% excess air	11.27	1	1.12	0.52	0.585	0.74
lb/1000 lb flue gas at 12% CO_2	10.0	0.89	1	0.46	0.52	0.66
gr/st ft³ at 50% excess air	21.31	1.93	2.16	1	1.12	1.42
gr/st ft³ at 12% CO_2	18.85	1.71	1.92	0.89	1	1.26
g/Nm³ at ntp, 7% CO_2	15.0	1.36	1.53	0.704	0.79	1

shows a variation from 10 lb to 60 lb of dust per
ton of refuse burned, and excluding the data from
some German incinerators where there was known to
be a high concentration of ash in the refuse fed to
the incinerator, with a mean value of 24 lb per ton.
The value may be compared to a figure of 17 lb per
ton reported by the Public Health Service,[3] 35 lb
per ton reported by Fernandes,[1] or mean values for
stack emissions for furnaces without special fly
ash collection devices of 20 lb per ton reported by
Rehm[4] and also by Chass and Rose.[5]

It is of interest to compare these data with the
recent work of Kaiser and Carotti[6] for the Society
of the Plastics Industry, done at the Babylon, Long
Island incinerator. For two experiments reported in
detail, the amount of particulates reported are 40
and 41 pounds per ton of refuse burned, somewhat
higher than the averages reported above, but generally
in line with them.

Dust sizing, like dust loading, varies widely.
The dust is quite heterogeneous, having a chemical
composition similar to that of the ash of the principal constituents of the refuse from which it came,
but usually including some large, low density flakes.
Dust density varies from an average of slightly over
2 grams per cubic centimeter (125 pounds per cubic
foot) to as high as 3 grams per cubic centimeter
(187 pounds per cubic foot). About 35% of the
average dust leaving the furnace is smaller than 10
microns, with the bulk of the particles in the range
of 10-250 microns. Simple settling chambers and
spray devices do not usually remove very small
particles sufficiently to meet the growing demands
for more stringent limits of emission.

It will be recognized that the 35 pounds per ton
mentioned above is a value before any deliberate
attempts have been made to remove the fly ash. The
question still remains as to how much reaches the
atmosphere and how much can be tolerated. Present
practice suggests conformance with the 1949 ASME
example for a smoke regulation ordinance of 0.85
pounds per 1000 pounds of flue gas, adjusted to 50%
excess air. This is approximately equal to 10 pounds
per ton of refuse burned, or, based on a 20-35 pound-
per-ton generation, a called-for reduction of only
50-70%. In a new suggestion by ASME in 1966, the
permitted figure would be 0.68 pounds per 1000
pounds of flue gas, adjusted to 50% excess air, or
7.7 pounds per ton of refuse burned. This would be
equivalent to a reduction of 60-80%. The city of

New York has a more stringent code requirement of
0.2 pounds per 1000 pounds of flue gas, and to meet
this requirement would mean a reduction of 88-94%
on the same basis as above. The state of New Jersey
(proposed Chapter XI) calls for the same low level.
    In the Federal Register of August 17, 1971,
standards of performance for new stationary sources
are given, with particular reference to incinerators.
It should be stated here that the new proposed
standard at the federal level for particulate emission is 0.01 grams/scf (0.23 grams/Nm^3) corrected
to 12% CO_2, maximum 2-hour average. This is
equivalent to about 0.02 pounds per 1000 pounds of
flue gas, which, if meant seriously, would be ten
times more severe than the requirements of the city
of New York or the state of New Jersey.
    Walker and Schmitz[7] give some data for other
cities in the United States as of several years ago
for maximum particulate emission for large refuse
incinerators:

Agency	*Gas Adjusted, Maximum Permitted*
Cincinnati, Ohio	0.36 lb/1000 lb
Detroit, Michigan	0.30 lb/1000 lb
San Francisco Bay Area, California	0.68 lb/1000 lb
Federal Guides	0.34 lb/1000 lb

    Walker and Schmitz[7] state that the attention
being focused on the general problem of air pollution
in the United States is leading to more stringent
regulations and enforcement of quantitative emission
standards. As a result, a greater number of air
pollution sources, which have heretofore not been
considered as significant contributors to the general
air pollution, are coming under close scrutiny. One
of these is the large, mechanically-stoked incinerator for the disposal of solid waste, ranging from
sewage sludge to domestic and commercial refuse to
demolition and construction debris. A combination
of circumstances--increasing population, decreasing
land availability, increased per capita generation
of refuse, prohibition of open burning, etc.--point
to incineration as a growing potential source of
air pollution if it is not properly controlled. Thus,
emission regulations governing this source are becoming increasingly strict. Where values of 0.85
pounds per 1000 pounds of flue gas corrected to 50%

excess (or 12% CO_2) were once the accepted rule, many local, state and regional codes now call for much lower maximum emission rates.

More recently, the Environmental Protection Agency has published in the *Federal Register* of August 17, 1971 the first set of proposed federal standards for performance of stationary sources of air pollution, including incinerators.

VISUAL EMISSION LEVELS

In addition to provisions limiting the quantities of particulate emissions, many municipal and state codes also have desired opacity restrictions, usually based on the so-called Ringelmann chart.[8]

The Ringelmann number is determined by a comparative visual observation of the stack plume and a series of reference grids of black lines on white that, when properly positioned, appear as shades of gray to the observer. Although the quality of a plume in equivalent Ringelmann numbers is not easy to determine, trained observers, properly positioned in relation to stack, sun and wind direction, can provide satisfactorily consistent evaluations. Water vapor plumes complicate observations of stack gases, but trained observers can distinguish between water vapor and residual plumes under selected weather conditions.

The trend in most codes on incinerator emissions based on Ringelmann is to require that (1) normal, continuous plume quality not exceed Ringelmann No. 1; (2) for short periods not exceeding 3 to 5 minutes in any one hour, plume quality not exceed Ringelmann No. 2 and (3) plume quality clearly resulting from water vapor only be excluded from regulation.

Plume opacity refers to the inability of light to pass through the gas plume and is usually applied in cases where the plume is some color other than gray or black. Opacity readings are expressed in terms of percent visibility through the gas plume.

Where water injection is used to reduce flue gas temperatures and particulate emission, the resultant gas leaving the stack will contain excess moisture and form "plumes." Two cases arise, one where only gas cooling is involved and the other where both functions are performed.

In Case 1, absolute humidities may be several times higher (approximately 0.25 to 0.30 lb water vapor/lb dry gas) since a substantial amount of water

is evaporated in cooling the gases from stack
temperature of 1,200 to 1,800°F down to 450 to 600°F;
however, stack temperatures are still high, dispersion capability is good and condensate plumes will
probably be limited to situations of low ambient
temperature and intermediate ambient temperatures
associated with high relative humidity. In Case 2,
particulate scrubbers saturate the gas with moisture,
absolute humidities are high (under adiabatic
saturation conditions approximately 0.6 lb water/
lb dry gas), and effluent temperatures are low (in
the range of 175 to 180°F under adiabatic saturation
conditions) so that dispersion capability is poor
and condensate plumes will occur under almost all
atmospheric conditions.
   Condensate plumes are usually not harmful, and
local complaints may possibly be reduced by proper
preeducation and public relations regarding water
vapor plumes. On the other hand, there have been
complaints of corrosion on automobiles resulting
from condensate fallout, and of a decrease in
visibility at ground level on roadways that have
been connected with water vapor plumes from incineration. Thus, these potential problems must be
given serious consideration in the design of the
plant.
   Experience with large, coal-fired steam generators
indicates that loadings in the range of 0.01 to 0.02
gram per cubic foot of exit gas result in stacks
optically clear of suspended particulates (*i.e.*, less
than Ringelmann No. 1). Achievement of these stack
concentrations in coal firing requires collector
efficiency in excess of 99% by weight. Similarities
in particle size distribution from coal-fired steam
generators and from incinerators indicate that
efficiencies of this order will be required on
incinerators if completely clear stack emissions
are to be achieved.

CONTROL OF ODORS

   The best approach to the control of odors generated
in the drying and combustion process is maintenance
of adequate retention time and sufficient temperature
to ensure complete combustion of hydrocarbon vapors
to carbon dioxide and water. Elimination of odors
from stack gases demand that mixing of any volume of
gas containing odors must be completed so that the
required excess air and temperature conditions are

reached in every stream of gas. A 0.5-second residence time within the combustion chamber, with a
temperature at the furnace exit port of 1400°F or
above, will be sufficient to eliminate odors.

Another odor control technique is dilution of
the odorous gas in the atmosphere to a value below
its threshold odor so it is unapparent to a receiver.
This is achieved through the use of stacks of sufficient effective height. Effective stack height is
a function of both the actual stack height and the
plume rise as the gases leave the stack. The height
to which a plume will rise above the stack is a
function of the ambient temperature and the gas
temperature, exit velocities of the stack gases,
and the stability of the atmosphere. If the
threshold value of any identifiable odorous gas is
known, methodology is available for estimating the
maximum quantitative emission that will keep odorous
gas below the threshold value at the nearest receiver.
Presently, however, this technique for controlling
odors is not being used for two reasons: (1) the
ability to eliminate odors by proper process operation; and (2) the absence of identifiable odorous
contaminants and their threshold values.

Data on 107 existing batch feed plants indicate
a maximum stack height (above grade) of 250 feet, a
minimum of 39 feet, and an average of 133 feet. On
44 continuous feed plants the values were: maximum
250 feet, minimum 25 feet, and an average of 145
feet. The advances that have been made in the use
of meteorological methods and data in the design of
stacks as pollutant dispersion devices appear
applicable to the design of stacks for incinerator
plants. This suggests the desirability of competent
meteorological consultation to determine stack
heights.[8]

CONTROL OF OTHER GASEOUS EMISSIONS

Both nitrogen oxide and sulfur oxide emissions
occur in solid waste incineration, but the amounts
per ton of fuel burned are several orders of magnitude below those involved in the combustion of fossil
fuel. Solid waste is inherently a "clean fuel" from
the standpoint of sulfur content, with a value of
about 0.16% by weight as compared to most coals and
residual oils used today which range from about 1
to 3% sulfur. Further, there is evidence to suggest
that the sulfur is mostly retained in the ash rather

than emitted as oxides in the stack. Thus, sulfur
oxide emissions from solid waste incineration
generally are well below even the most stringent
restrictions, present or anticipated.[8]

Nitrogen oxide emissions per ton of fossil fuel
are over ten times greater than those from an incinerator. Therefore, it appears unlikely that this
contaminant will be regulated in incinerator plants
in the near future.

Formation of nitrogen oxides from the incineration of refuse is believed to be a function of
temperature in the incinerator and not a result of
any ingredient in the refuse *per se*. Thus hot spot
temperatures in excess of normal, or operation with
an input refuse of higher than normal calorific
value and inability to control furnace temperatures
below 2000°F will lead to an increase in the amount
of nitrogen oxides produced.

Where gas scrubbing is employed, both the sulfur
and the nitrogen oxides will be decreased, especially
when alkaline liquors are used.

The European countries, in general, have more
rigorous official standards and have been operating
successfully under them for some period of time.
The more widespread use of electrostatic precipitators,
for example, is undoubtedly due to lower tolerance
for particulate emissions. It is interesting to compare data of European practice with those of the
United States, and this has been done by Rogus.[9]
Thus, on the subject of particulates, Germany and
Austria have a limit of 0.21 lb/1000 lb flue gas,
France 0.11, Italy and Spain 0.85, and the United
Kingdom and Denmark more stringent criteria than
Germany. While the Italian and Spanish standards
are equivalent to the ASME 1949 standards, those
for the more highly industrialized countries are
more stringent than normal U.S. limits, with New
York City and New Jersey coming to the same low
levels recently.

It should be noted also that European practice
favors much higher stack height than in the United
States (300-600 ft) so as to promote better natural
dispersion of both particulates and gases.

The European feeling is that American practice
is less stringent and no American refuse incinerator
built before 1969 has been found to comply with even
the lower standards when subjected to careful tests.

It is worth noting here that present testing
methods are time consuming and require considerable
skill to assure dependability of results. Accordingly,

such testing is infrequent and of historical significance only. To correct this situation, suitable
instrumentation is being developed under the auspices
of the American Public Works Association and the
Public Health Service. These instruments will provide prompt readings and recordings to guide incinerator operation and to furnish official daily
records.

AIR POLLUTION CONTROL
EQUIPMENT PERFORMANCE

Furnace Emission Collection

As discussed by Fernandes,[1] particulate collection
equipment performance may be classified in a number
of ways, but the most widely accepted criterion is
the weight efficiency.
The weight efficiency relates the quantity of
the dust collected to the dust that enters the collector with the gas. This number may be meaningfully
applied only under conditions similar to those existing
during the test, including the given dust density and
size distribution, the entering gas dust loading, the
collector energy level, and the inlet gas temperature.
The results can sometimes be related to other applications if the dust density, size distribution, dust
resistivity (if a precipitator is used), collector
energy level, and gas condition are known.
Another important collector performance criterion
is the fractional efficiency curve, sometimes called
the size or grade efficiency curve. It represents
the performance of the particular collector on each
size of dust particle of a given density for a given
collector energy level, gas temperature, and dust
resistivity (if a precipitator). The two efficiencies
are related and can be computed one from the other if
the dust size distribution is known. This is very
important since most air pollution control equipment
manufacturers would prefer to guarantee the known
fractional efficiency performance for their equipment
and allow each purchaser to compute the efficiency
for his particular dust.
The size and composition of incinerator fly ash
and the extremely large quantities of air used in
incineration mask the real pollution potential. As
a result, stack observations are no measure of an
incinerator's pollution control. An accurate determination of stack emissions can be obtained only by

actual tests based on samples taken in the duct
leaving the air pollution control equipment. It is
suggested that test connections be designed into
the ducting before and after the primary dust collection equipment. A makeshift arrangement to
accommodate sampling at a later date is, at best,
a compromise. Pre-engineered sampling points will
permit accurate measurement of particulate emission
from the stack and testing of the primary air
pollution control equipment to determine if it is
functioning properly.[1]

Collector performance and the resulting emission
to the atmosphere are summarized in Figure 9.1.
These can be compared to the local emission standards
which may be used as an entry to the graph.   The

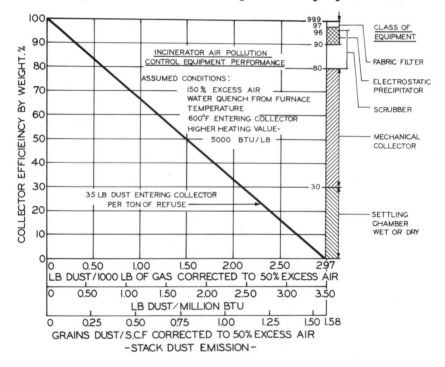

Figure 9.1. Collector efficiency versus stack dust emissions.[1]

efficiency required is read on the left ordinate while
the right ordinate presents the class of air pollution
control equipment that could be designed to meet this
requirement. As an example, if the ASME 1966 maximum
emission level   is used, one can start with the 0.8 lb

of dust per million BTU and read 77% efficiency on
the left ordinate and on the right ordinate note
that a mechanical collector could be designed for
this service. Once again, the reader is cautioned
that these data assume a properly designed and maintained collector and an incinerator with good combustion conditions. The ranges of performance
presented on the right ordinate of the figure
indicate areas in which it is reasonable to expect
each class of equipment to perform. In most cases,
the 35 lb of dust per ton of refuse leaving the
furnace, assumed as a basis for this graph, is a
satisfactory starting point. If, for a certain type
of incinerator, the designer knows the furnace
emission is greater or less than the assumed 35 lb
per ton, a second line can be drawn (from the 100%
efficiency and zero emission point to the expected
furnace emission on the zero efficiency line) and
the graph used as before for these new conditions.

Figure 9.1 illustrates that today's technology
can provide good pollution control on modern incinerators. As mentioned earlier, the highest efficiency
collectors may require additional development to
achieve their full potential, but they are available
to the industry today.[1]

Particulate Efficiency Requirements
for Municipal Incinerators

Walker and Schmitz[7] state that their basic objective has been to establish some realistic requirements
for particulate cleaning efficiency in municipal
incinerators to meet existing and future quantitative
emission codes.

To their knowledge, the information presented in
the paper, including the Stenberg reference, constitutes the sum total of such published data on modern
American incineration practice. The tests on three
different grate configurations in three distinctly
different municipal refuse incineration operations--
large urban northeastern U.S., small northeastern
U.S., and medium-size suburban midwest--constitute
the best available representative sample of this
practice. They should provide a reasonable basis
for some generalizations on particulate removal
efficiencies required between furnace outlet and
stack to meet various maximum emission criteria.
In view of the apparent absence of any significant
relationship between type of grate and particulate
emission, the inability at the present time to predict

with any certainty the composition of the fuel and
its relationship to particulate emission, and the
wide variation in individual operator technique with
regard to underfire air, the only reasonable approach
appears to be to treat all of the available measurements of furnace emissions as a finite population to
establish these levels of cleaning efficiency.

There are essentially two classes of criteria
which must be of concern: quantitative emission
criteria and optical opacity criteria. Quantitative
emission criteria are usually established on the
basis of grains of particulate per standard cubic
foot of dry flue gas corrected to some reference
level of CO_2--usually 12%--or on the basis of pound
of particulate per 1000 pounds of wet gas corrected
to 50% excess air.

Figure 9.2 shows minimum particulate removal
efficiencies required with a reasonable assurance
that one will meet the required code 95% of the time.

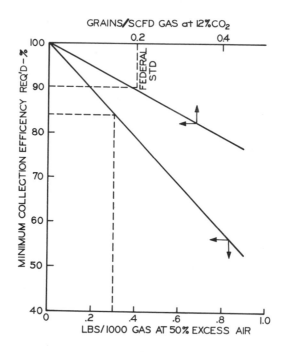

Figure 9.2.   *Efficiencies required from the furnace outlet to
meet various levels of quantitative emission.*[7]

For example, to meet the federal guides governing emissions from incinerators of 0.2 grans/SCFD at 12% CO_2, one requires a removal of 90% of the particulate leaving the furnace. Similarly, to meet the Detroit code of 0.3 lb/1000 lb gas at 50% excess air, one would require a minimum collection efficiency of 84%.

Considering optical criteria, the rather similar particle size characteristics of the sub-sieve fraction of incinerator ash and fly ash from pulverized fuel boilers allows us to utilize particulate emission criteria for optically clean stacks which are fairly well established in the power generation field. The usual criterion for a clear stack on a power boiler is 0.01 to 0.015 grains/SCFD. Considering the differences in combustion practice of pulverized coal (at stoichiometric to 50% excess air and 300°F at the stack) and present municipal incineration practice (approximately 300% excess air and 600°F at the stack), these values would be equivalent to approximately 0.03 to 0.05 grains/SCFD. Based upon a furnace emission of 0.76 grains/SCFD, a value below which 95% of the test measurements fell, an efficiency of 93.5% will give a reasonable assurance that the stack will be optically clear of particulate.[7]

Factors in the Selection
of Particulate Collectors

Matching the type of collection equipment needed to meet a particular air pollution control objective is illustrated in Tables 9.3 and 9.4.

Table 9.3

Collection Efficiency Required to Meet
Various Emission Limitations[8]

Code Requirement *(lb particulate/1000 lb flue gas)*	*Approximate % Efficiency*[a] *to meet Code*
0.85/1000 at 50% excess air	74
0.65/1000 at 50% excess air	80
0.20/1000 at 50% excess air	94

[a]Based on 32 lb of fly ash per ton of solid waste charged entering the collector.

Table 9.4

Maximum Demonstrated Capability of Various Collectors[8]

Type of Collector	Maximum Demonstrated Efficiency (%)
Settling chambers	34
Wetted baffles	53
Cyclone collectors	70-80
Direct impaction scrubbers (wet scrubbers)	94-96
Electrostatic precipitators	99
Bag filters	99+

Unless furnace operation is to be significantly restricted, particularly with respect to underfire air, even the most lenient quantitative emission code cannot be met with settling chambers or wetted baffles, either alone or in combination. Cyclone collectors can meet 0.85 lb per 1000 lb at 50% excess air and probably can meet intermediate codes. When codes require emissions below 0.65 lb per 1000 lb at 50% excess air, the only demonstrated alternatives are direct impaction scrubbers, electrostatic precipitators, or bag filters.[8]

REFERENCES

1. Fernandes, J. H. "Incineration Air Pollution Control," paper prepared for National Incinerator Conference (ASME) at New York (May 5-8, 1968).
2. Niessen, W. R. and A. F. Serofim. "Incinerator Air Pollution Facts and Speculations," paper prepared for National Incinerator Conference (ASME) at Cincinnati, Ohio (May 17-20, 1970).
3. Duprey, R. L. "Compilation of Air Pollutant Emission Factors," National Center for Air Pollution Control, Environmental Protection Agency, Public Health Service Publication No. 999-AP-42 (1968).
4. Jens, W. and F. R. Rehm. "Municipal Incineration and Air Pollution Control," paper prepared for National Incinerator Conference (ASME) at New York (May 1-4, 1966).
5. Chass, R. L. and A. H. Rose. "Discharge from Municipal Incinerators," *Air Report 3(2)* (1953), pp. 119-122.
6. Kaiser, E. R. and A. A. Carotti. "Municipal Incineration of Refuse with 2% and 4% Additions of Four Plastics," report to Society of Plastics Industry, New York (June, 1971).

7. Walker, A. B. and F. W. Schmitz. "Characteristics of Furnace Emissions from Large, Mechanically-Stoked Municipal Incinerators," paper prepared for National Incinerator Conference (ASME) at New York (May 1-4, 1966).

8. DeMarco, J., D. J. Keller, J. Leckman, and J. L. Newton. "Incinerator Guidelines," Public Health Service Publication No. 2012, Office of Solid Wastes Management Program, Environmental Protection Agency (1969).

9. Rogus, C. A. "An Appraisal of Refuse Incineration in Western Europe," paper prepared for National Incinerator Conference (ASME) at New York (May 1-4, 1966).

CHAPTER 10

RECOVERY OF HEAT AS STEAM AND
SALVAGE OF RESIDUES

HEAT RECOVERY AND UTILIZATION

The recovery and use of heat produced during incineration has been discussed by DeMarco, Keller, Leckman and Newton[1] who point out that in European municipal incinerators, in contrast to those in the United States, the use of heat recovery equipment is quite common. Heat recovery and water cooling in incinerator furnaces has the added advantage of reducing the volume of gas to be cleaned and the amount of particulates entering either the gas-cleaning equipment or the stack itself. With the increasing demand for greater efficiency in gas cleaning, the justification for installing heat recovery equipment improves but is still highly dependent on the ability to utilize the recovered heat in an economic fashion.

Four basic designs of heat transfer have been used: waste-heat boiler systems in which tubes are located beyond conventionally built refractory combustion chambers; water tube wall combustion chambers; combination water tube wall and refractory combustion chambers; integrally constructed boiler and water tube wall construction.

Refractory-lined chambers usually require 150 to 200% excess air to maintain temperatures, while water tube wall chambers can operate at 50 to 100% excess air. In the absence of heat recovery equipment, large quantities of cooling spray water or additional cool air are required to reduce the gas temperatures to the 600°F range required in most air pollution equipment. Steam production ranges from 1 to as much as 3.5 pounds per pound of refuse incinerated; heat recovery may be as great as 70%, depending on the composition of the refuse at the time of burning.

In heat recovery systems provisions must be made to keep the boiler tubes clean and to prevent ash buildup, slagging, and corrosion of the boiler tubes, particularly on the fire side. Increasing the space between the water tubes can reduce fly ash clogging of gas passages. External corrosion of boiler tubes on the fire side of the tubes can result in high maintenance costs. Most of the reluctance of U.S. engineers in designing waste-heat recovery systems is based on added cost, variability of the heat content of the refuse, and difficulty in matching the supply of such waste heat to the demand for heat. An incinerator's daily quota of refuse displays varying heat content and, in addition, requires periodic delays for maintenance and overhaul. Output of steam is therefore inconsistent, and provisions must be made to either provide or dissipate additional heat. Excess steam can be condensed by heat exchangers and recovered water recycled, while additional steam can be provided by auxiliary fuel burners.

Many U.S. incinerators equipped with heat recovery systems use such heat for in-plant use only, such as electricity generation, hot water supply and heating the building. The siting of an incinerator to insure a market for any recovered heat is an obvious advantage; examples are institutions such as hospitals, industrial parks, and high-rise apartment buildings. Sale of steam to power generation plants is another possibility.

Aside from resource conservation and air pollution control factors, the decision to practice heat recovery at an incinerator site is based primarily on economic considerations. The sale of steam requires that the incinerator site be close to the buyer because of the high cost of distribution. Assuming steam production of 1 to 3.5 pounds per pound of refuse incinerated, at 1969 prices for steam, DeMarco *et al.*[1] estimate returns of $0.50 to $3.50 per ton of refuse burned. This is based on a 10 mils per kilowatt hour with a steam power cycle efficiency of 25% yielding 7,300 kwh per 1000 pounds steam.[1,2] Stephenson and Cafeiro,[3] in a study involving reported data from 205 municipal incinerators, show that only 43 such incinerators reported waste heat utilization. Data in Table 10.1 show that 24 plants used the waste heat for in-house heating, for hot water or both; 9 plants used the heat in hearby sewage plants; 5 produced electric power; 2 sold steam outside; the remainder used waste heat for preheating combustion air or for unspecified purposes.

Table 10.1

Waste Heat Utilization[3]

| Period | Plants Reporting Use | Bldg. Heat and/or Hot Water | Electric Power | Sewage Sludge Drying | Steam Production | | | | Preheat Combustion Air | Other Use |
					For Sale	Outside Heating	Other Use	Use Not Stated		
1945–1950	2	1	1	0	0	0	0	0	0	0
1951–1955	10[a]	4	2	2	1	0	0	2	1	0
1956–1960	17[b]	10	1	3	0	1	d,g	1	1	e,f
1961 to date	14[c]	9	1	1	1	1	h	1	0	i
TOTALS	43	24	5	6	2	2	3	4	2	3

Number of Plants Using Heat For:

[a] One plant reports building heat, hot water, and preheating combustion air; another reports building heat and sludge drying.

[b] One plant each report hot water and power generation; hot water and air preheating; hot water and sludge drying; and steam for equipment drives and heating nearby hospital.

[c] One plant reports building heat and steam for sale; one reports power generation and desalination.

[d] Equipment drives.

[e] Sludge furnace.

[f] Heat for sludge digester.

[g] For sewage treatment plant.

[h] Desalination of sea water.

[i] Tubular gas reheater cools combustion-chamber outlet gas and reheats scrubber exit gas.

Rogus,[4] in a review of European versus U.S. practice, points out that today's potential for improving incinerator performance presents a more favorable climate for waste heat recovery because:

- The heat content of refuse has increased about 50% in the last decade. Further substantial increases are predicted, particularly if the high-heat over-sized burnables are crushed and added to normal incinerator refuse.
- More efficient burning is now possible through substantially improved grate and furnace construction. These improvements now achieve an 80 to 85% burning efficiency and are capable of handling refuse at rates of 85 to 100 pounds per square foot of grate surface per hour.
- Costly refractories can now be replaced with maintenance-free water-cooled walls which should reduce the size and cost of separate boiler construction and its housing, provide a more efficient direct heat absorption and utilization, and produce an efficient cooling system for the raw gases so as to reduce substantially the capital and operating costs of air pollution control equipment.

A brief review of European experience and current practices may be helpful in revising past thinking. In post-war Europe, economic unavailability of rich fossil fuels stimulated research and development for substitutes. Progressively the Europeans improved the design of refuse incinerator and waste heat recovery systems so that practically all of their new large incinerators are equipped with steam and/or power generating mechanisms and include many relatively long transmission conduits. Some of these new plants, as in Paris, Rotterdam and Vienna, limit themselves to the burning of refuse. Others, as in Munich, Stuttgart, and Hamburg, supplement refuse with pulverized coal in more or less equal amounts.

In typical large European incinerators, the furnace combustion chamber walls and the rear refractory arch are lined with tubing which circulates water at 400 to 450°F. The high flames and hot gases are directed vertically for optimum combustion and heat absorption. Near the top the flow of gas is reversed over a bridge wall into a second boiler pass and then, near its bottom, is turned upward again to enter the third pass housing the superheater. Third and fourth reverses circulate the gas downward and then upward in the two banks of a convector section before entering the horizontal economizer pipe bundles and the final discharge ports. The tortuous gas flows help to precipitate out as much as 50% of the particulate matter to the bottoms of the four hoppers. This flyash is transported by gravity or pneumatically into the quenched residue system. Tube spacings in the furnace chamber are 1 to 2 inches in size, but beyond the furnace they must be 3 to 4 inches to alleviate clogging. Low gas velocities are required to avoid abrasive action. Corrosion is held to a minimum by maintaining temperatures above the dew point. Cleaning of tube surfaces to free them of flyash may be done by continuously recirculating and dropping steel pellets from an overhead distributing hopper; however, retractible soot blowers, if available, do a better job of flyash removal. Tubes are more easily cleaned if placed parallel to the gas stream flow.[4]

European operating records show that although one pound of their incinerator refuse can produce up to 1.8 pounds of steam, a dependable average approximates 1.4 pounds released at a rate of 2.5 pounds of steam per square foot of boiler surface. However, since at these rates only about 4% of the refuse-producing population could be provided with heat at peak winter loads, supplementary fuels and heating plants are installed to accommodate such peak demands.

In Europe, central heating and appropriate distribution systems are used extensively. It is reported that in West Germany alone (1962) there were 75 district heating plants having an installed load of 4 billion BTU per hour and a pipe line distribution system totaling 3,300 miles. The major advantages of such systems over individual heating plants are savings in construction and operating costs, savings in fuel distribution, and substantial reductions in air pollution through more efficient burning, use of sophisticated air pollution abatement apparatus and super-high (600-foot) stacks.[4]

Where the connected summer loads, generally about 10% of the median winter requirements for steam, are substantially below the incinerator's actual output, the surplus steam is converted to electrical power. On the average, 10 pounds of steam is required per kilowatt-hour. It should be noted that of the generated steam and electricity about 10% are required for on-site plant use, at a rate of about 30 kwh per ton of refuse.

As has been mentioned earlier, there are several installations in the U.S. utilizing the waste heat generated from refuse incineration. Day and Zimmerman Associates[5] report that most of these

employ conventionally designed steam boilers with
relatively close tube spacing to obtain maximum
steam capacity and efficiency at minimum first cost.
Chicago's new Northwest incinerator is the first
U.S. installation to use (a) water-walled furnace,
(b) steam generation and (c) electrostatic precipitation. With its 1600-ton-per-day capacity, it uses
generated steam to drive the major plant auxiliaries
such as combustion air and induced draft fans, boiler
feed pumps and refuse shredders. Studies indicated
that, under the conditions in Chicago, it was more
economical to use electricity from the local utility
than for the city to use the incinerator-generated
steam to operate a small power plant.

Additional information bearing on the question
of waste heat recovery from incinerators may be found
in the following:

Cohan and Fernandes[2] point out that the control
of off-gases, wastewater and their contaminants can
be accomplished through process engineering applications. Some of the prospects enumerated are:
regenerative feed water heating, district heating,
district air conditioning, refrigeration, desalination, separately fired superheaters, and incinerator
gas turbines. Solutions are complex but not
insurmountable.

Regan[6] discusses the generation of steam from
prepared refuse. Refuse is reduced, by methods
described, to about 2 x 2 inches before being
pneumatically conveyed either directly to a furnace
or to a storage area. One such incineration system
described is a 180 ton-per-day saturated steam-
producing, 100% refuse-and-sludge-fired unit.
Refuse throughput is 180 tons per day, and sludge
is dried and incinerated at the rate of 114 tons
per day; the steam flow is 77,000 pounds per hour
at 400 psig. A second installation involves a combination refuse and oil-fired unit--the refuse
generating about one-third of the steam flow of
300,000 pounds per hour at 825°F and 650 psig. The
third unit mentioned is a coal-fired utility boiler
adapted to burn 10% refuse--approximately 250 tons
per day of 6000 BTU per pound of refuse. The
capability of improving methods of conveying and
burning refuse exists to have better burnout, reduced stack emission, and reduced cost of refuse
conveyance, residue removal and flue gas cleaning.

The steam producing facilities of the Mannheim-
North incinerator in Germany are described by
Hilsheimer.[7] The decision to build an industrial
development on the site of the existing sanitary

landfill led to the decision to build an incinerator as part of a power plant complex to service the development. The population of Mannheim is approximately 335,000, generating about 160,000 tons of refuse per year (450 tons per day) of average calorific value, 2,700 BTU/lb and 36% moisture content. Two traveling grate furnaces are provided with a water wall furnace enclosure and a natural circulation boiler with auxiliary oil firing. The heating areas consist of a radiation section with wall superheater, a convector section with final superheater, a convection superheater, water and air preheater. Gas cleansing is by electrostatic precipitation. The units were designed for the following operating conditions:

Steam production (each boiler)	40 tons/hour
Steam pressure--design	136 atm
Steam pressure at boiler discharge	119 atm
Steam temperature	525°C (977°F)
Makeup water entrance temperature	160°C (320°F)
Refuse burning rate (each unit)	
Average	15 tons/hour
Maximum	18 tons/hour
Heat from refuse	27 Gcal/hour

Stephenson and Cafeiro,[3] in an excellent survey of incinerator design practices and trends, show the improvements and changes in incinerator technology over the past twenty years. Their report on the findings of a survey of design practices covering 205 plants built or designed since 1945 lists 43 plants in which waste heat recovery is practiced and gives their locations plus other important data on operations (see Table 10.1).

Eberhardt[8] indicates that American and European incineration practices emanate from two different prerequisites. In the U.S., reduction of the refuse is the primary goal. In Europe, the aim is to completely burn out the refuse, utilize the waste heat, and minimize air pollution through the use of expensive flue-gas cleaning equipment. Descriptions of refuse boilers and their construction, grate construction and firing systems, combined oil-refuse firing systems, emissions and control thereof are given. Economics are considered, and problems of sewage sludge disposal are treated.

Eberhardt and Mayer[9] consider that European steam generators with refuse firing must meet a number of stringent legal requirements for environmental control. Flue dust collectors have 98% or

better efficiency. Physical and chemical problems
with the fuel and with boiler availability are met
by attention to specific engineering details. Corrosion of boiler and superheater tubes is largely
prevented by maintaining oxidizing conditions in
critical areas. Covered in the text are: air
pollution control, water pollution control, an
excellent section on furnace and boiler design,
boiler fouling, and fire-side tube deterioration.
     Stabenow[10] discusses design and performance
criteria for the layout of large municipal incinerators. The study points out that a formal code for
such design and performance, although needed, is
still in its preliminary stage. The value of a
variety of designs from a wide range of municipal
incinerators which have operated successfully in
Europe over long periods is covered. The heat
recovered through generation of steam permits a
thorough evaluation of incinerator performance
tests; as a result, it should now be possible to
prepare performance specifications for high grade
incinerator equipment. Enforcement of such specifications should permit municipalities to select
incinerators that will assure a long life and
satisfactory operation. Stabenow points out some
of the design and performance specifications for
incinerators with heat recovery. For example,
suggested specifications are considered for:
intake hopper, up to and including the induced
draft fan connection to the stack; feed chute;
grate and air plenum chamber; ash and clinker removal devices; the furnace and boiler for heat
utilization (type of construction, heat release,
spacing of screen tubes, boiler design, gas velocity);
the convection section; the economizer; and cleaning
of the boiler heating surface. Also it is mentioned
that an ASME Performance Test Code Committee is preparing a procedure to test large incinerators.
Such a code will enable the municipal authorities
to accept or reject an incinerator installation
after completion of a series of tests.
     A number of other articles make contributions
to the art of incinerator design for heat recovery,
but the above should give the interested reader some
background information pertaining to the subject.

SALVAGE OF INCINERATOR RESIDUES

     In a study of resource recovery from incinerator
residue,[11] the American Public Works Association

(APWA) analyzed factors that affect recycling of
ferrous metals and other inorganic material contained
in municipal incinerator residue.

Nonferrous Metals

Refining tests have been conducted by the U.S.
Bureau of Mines on a limited number of mixed nonferrous metals obtained from incinerator residues.
Table 10.2 contains the results of a test using
vacuum distillation to separate zinc and lead from
the aluminum matrix. This cannot be considered
typical of all such products, but it gives an indication of the possibilities of separating fractions
that would be acceptable to secondary metals
smelters.[11]

Table 10.2
Nonferrous Metals Contained in Incinerator Residues[11]

Constituent	Head Sample (%)	Condensate (%)	Aluminum Heel (%)
Al	61.0	0.14	83.6
Zn	19.7	Balance	0.05
Cu	1.8	0.01	2.43
Fe	1.4	0.01	1.62
Pb	1.4	1.53	0.56
Si	1.0	-	1.30
Balance	Probably aluminum oxides and minor impurities	-	Probably aluminum oxides and minor impurities

Tin may reasonably be expected to be present in
any nonferrous product separated from residues. It
should normally be less than 1 per cent, but always
in objectionable amounts. It is not removed by
vacuum distillation.

Ferrous Metals

Market outlets for ferrous metals contained in
incinerator residues are uniformly reported to be
nonexistent with the exception of the mining industry.
The availability of high-grade scrap precludes

marketability of low-grade scrap, and, were the
ferrous metal fraction of incinerator residues cleaned
and refined to meet present minimum requirements for
steel production, the impact of this limited quantity
of scrap on the existing local market would not be
significant or might depress the market. There is
a limited outlet for processed shredded ferrous metal
for copper precipitation, a market which lies mainly
in the southwest quadrant of the United States. It
is estimated that this outlet receives about 400,000
tons of scrap annually from all sources.

The relative importance of the ferrous metal
fraction of incinerator residue can hardly be over-
emphasized. Analyses of combined domestic refuse
processed through high temperature incinerators
indicate the ferrous fraction approximates 28% of
the residue on a dry weight basis. The deleterious
effects of tin are consistently advanced as the
principal reason why tin scrap is generally unacceptable to steel mills and foundries. Markets for
baled, bright tin cans are therefore nonexistent.
Under present conditions, burnt (incinerated) tin
cans are said to be useless as scrap if they contain
more than 0.003% tin. This level may be practically
or economically unattainable since high temperatures
result in the formation of iron-tin alloys that are
extremely refractory to conventional detinning processes. It has been forecast that by 1976
approximately 50% of all cans will be made from
tin-free steels; this development could alleviate
the tin contamination problem.

In a recent study by the Midwest Research
Institute it was reported that in the United States,
6,794,000 tons of steel are annually converted to
packaging and closures.

Glass

There is a demand for cullet in excess of
present supplies (1970). However, the amount of
glass in community solid waste greatly exceeds the
amount sought by glass manufacturing plants in each
city. Without exception, nine glass manufacturing
officials expressed interest in any program of
resource recovery in this area. Because the demand
for cullet presently exceeds the available supply,
it might be concluded that an intensive cullet
recovery program from incinerator residue and
community refuse is in order. Recent analyses by
the U.S. Bureau of Mines indicate glass makes up

about 44% by weight and 10% by volume of incinerator residue. However, cullet recovery has not been intensified for a number of reasons. Should economical and practical separation methods be developed to recover glass through an intensive solid waste salvage program, <u>including</u> incinerator residue salvage, supply would far exceed demand.

INCINERATOR RESIDUE PROCESSING

There is substantial interest in the principle of resource recovery among public and private solid waste management officials and industry managers, and a significant amount of recycle-reclamation activity is presently being practiced. It involves, however, only a small fraction of the solid wastes generated in a city. Also, some process of segregation and "cleaning" of salvageable items is fundamental to practical reclamation systems management, but such segregation will not in itself assure an acceptable recycling program of solid wastes, particularly incinerator residues. Finally, if resource recovery is to become an effective method of solid waste collection and disposal, market outlets emerge as the key factor.

Substantial quantities of commercial and industrial wastes are reclaimed through segregation at the source of waste production. Recycle of metal turnings through scrap processors is one example; recycle of self-generated cullet wastes by glass manufacturing plants is another. On the other hand, household refuse is collected as "mixed" or "combined" material and, therefore, presents a heterogeneous mass embodying significant economic and technological separation challenges. Assuming that separation is economically and technologically feasible, there remains the factor of contamination, especially cross contamination, *i.e.*, from product design mixtures such as tin cans with aluminum tops, impregnated rubber, paper products, etc. Serious contamination occurs from intermingled, adsorbed, or absorbed organic liquids and solids (food wastes, ink, oils, solvents, etc.).

Almost all cities use combined household refuse collection systems in conjunction with incinerators or sanitary landfills. Incineration offers a potential *zero residue* endpoint if resource recovery is practiced. Such a program envisions the removal of ferrous and nonferrous metals, recovery of glass, and refining of the remaining fly ash and residue to

recover trace metals or other values. The remaining
inert residue could then be used for building materials or as fill. The technology required is presently
the major thrust of the U.S. Bureau of Mines in
application and development work.

Data obtained confirm that development of market
outlets, whether economical or not, is fundamental
to the success of expanded resource recovery programs.
In other words, acceptance of locally generated items
contained in community solid wastes, such as ferrous
metals and glass, must be guaranteed on a long-range
basis. Despite the "combined material" approach in
packaging, aluminum recovery prospects appear to be
feasible under the incinerator residue processing
approach. While full cooperation of the glass con-
tainer industry can be anticipated in the use of
cullet, findings indicate that potentially salvageable
amounts would swamp this industry. Alternative uses
and new product approaches such as Glascrete will be
needed. The type of cullet produced in incinerator
residue processing is finely granular. The use of
shredded ferrous metal for copper precipitation purposes is a very limited outlet and new markets are
an absolute necessity. From a resource recovery and
recycling standpoint, market outlets mean profitable
outlets. From a refuse disposal standpoint, outlets
for recoverable items must be assured on a firm basis
for long-range periods before the necessary equipment
will be installed. Subsidies that stimulate the
profit motive do not necessarily provide outlets,
although they might aid to some extent in their
generation.[11]

Brackett[12] shows data on ash collected, ash
utilized, comparative utilization 1966-1969, and the
ash utilization trend from the burning of fossil
fuels. Fly ash generated from refuse incineration
is expected to have a different composition compared
to that generated from the burning of coal or oil,
but methods of collection and uses should be similar.
Yields of fly ash, of course, depend on type of
refuse burned, process and other factors, but 10-35
pounds per ton of refuse burned has been generally
reported. Assuming U.S. incinerator capacity at
about 55 million tons per year, fly ash yield
approximates 100 to 350,000 tons per year, or about
1% of the present yield from fossil fuels in power
plants.

Brackett states that the amount of ash shown in
Category 3 of Table 10.3 represents a high percentage
(about 25%) of the total ash utilized, and its end
use is not known. This is due to the fact that

Table 10.3

Preliminary Ash Collection and Utilization Survey* (1969)[12]

	Fly Ash (tons)	Bottom Ash (tons)	Boiler Slag (if separated from bottom ash) (tons)
1. Ash collected	22,304,513	3,042,017	3,020,282
2. Ash utilized:			
a. Mixed with cement clinker or cement (pozzolan cement)	32,759	–	–
b. Mixed with raw material before forming cement clinker	132,226	5,000	–
c. Partial replacement of cement in:			
(1) concrete products	175,086	5,686	–
(2) structural concrete	194,322	–	–
(3) dams and other mass concrete	163,809	–	–
d. Stabilizer for road bases, parking areas, etc.	264,224	20,617	73,240
e. Lightweight aggregate	250,251	19,388	–
f. Fill material for roads, construction sites, etc.	73,440	1,102,240	548,606
g. Filler in asphalt mix	114,067	10,084	89,970
h. Miscellaneous	244,508	648,052	257,204
Total	1,644,692	1,811,067	969,020
3. Ash moved from plant sites at no cost to utility but not covered in categories under "Ash Utilized"	452,522	381,076	48,387
Total Utilized	2,097,214	2,192,143	1,017,407

*Preliminary Figures--subject to change as tabulations are completed.

Table 10.4

Miscellaneous Uses for Ash* (1969)[12]

	Fly Ash (tons)	Bottom Ash (tons)	Boiler Slag (tons)
Abrasive cleaning	34	–	19,344
Test caps	3	–	–
Refractory add-mix	80	–	–
Experimental: ready mix supplier	62	–	–
Insulating cement	702	–	–
Oil well cementing	81,427	–	–
Grouting	3,832	–	–
Industrial tests	150	–	–
Snow sanding	800	–	–
Mine fire control	60,772	–	–
Unclassified	5,983	1,620	26,266
Subsidence control	31,750	–	–
Pipe coating	1,445	–	–
Foundry sand	1,550	–	–
Chemical products	250	–	–
Commercial ready mix	2,143	–	–
Foundries: manufactured products	2,300	–	–
Spontaneous combustion control	32,000	32,000	–
Oil well drilling	1,570	–	–
Highway bridges	11,361	–	–
Sewage treatment plant	3,000	–	–
Subsurface; heavy construction	2,819	–	–
Poz-O-Pac	475	–	–
Vanadium content	–	15	273
Ice control	–	441,039	34,567
School use on outdoor tracks	–	1,000	–
Asphalt shingles	–	108,877	5,033
Sandblasting grit	–	38,902	128,843
Dike repair and buildup	–	21,813	–
Drainage filter	–	2,786	200
Aggregate	–	–	42,678
TOTAL	244,508	648,052	257,204

*Preliminary Figures--subject to change as tabulations are completed.

several utilities sell their total ash production to
separate sales firms who, for proprietary reasons,
do not divulge how or where it is ultimately used.
    The wide variety of uses for ash is shown in
Table 10.4. In a sense, this diversity of use is an
area in which imaginative work is being performed
and the markets of tomorrow are being developed.
A few of the miscellaneous applications of bottom
ash and boiler slag, such as uses for ice control,
asphalt shingles, sandblasting or cleaning material,
and landfill, are moving appreciable quantities of
ash. In addition, fair amounts of fly ash are being
used for oil well grouting, mine subsidence, and
fire control. There is probably a certain amount
of overlapping in the applications of ash shown in
Table 10.4. Thirty miscellaneous uses for ash are
listed, and it is probable that there are many other
experimental uses for ash under investigation.

    While incinerator residue processing offers an
avenue for resource recycling as an integral part of
community solid waste management systems, there is
a fundamental need for research and development
geared to the generation of new products and uses
for low-grade ferrous metals and particulated glass.
More efficient utilization of the nonferrous portion
of incinerator residues depends in large part on
economic application of metallurgical refining
methods. The ultimate success of resource recycling
from incinerator residue is keyed to the expansion
of present markets and to the creation of new outlets
for recoverable items.

REFERENCES

1. DeMarco, J., D. J. Keller, J. Leckman, and J. L. Newton.
    "Incinerator Guidelines," Public Health Service Publica-
    tion No. 2012, Office of Solid Wastes Management Program,
    Environmental Protection Agency (1969).
2. Cohan, L. J. and J. H. Fernandes. "Potential Energy
    Conversion Aspects of Refuse," ASME paper No. 67 WA/PID-6,
    presented at Winter Meeting, Pittsburgh, Pa. (November,
    1967).
3. Stephenson, J. W. and A. S. Cafeiro. "Municipal Incinerator
    Design Practices and Trends," paper prepared for 1966
    National Incinerator Conference (ASME) at New York (May,
    1966).
4. Rogus, C. A. "Recent Trends of Refuse Collection and
    Disposal in Western Europe," paper prepared for 1966

National Incinerator Conference (ASME) at New York
(May, 1966).

5. Day and Zimmerman Associates. "Special Studies for
    Incinerators in Washington, D.C.," Public Health Service
    Publication No. 1748, Office of Solid Waste Management
    Program, Environmental Protection Agency.

6. Regan, J. W. "Generating Steam from Prepared Refuse,"
    paper prepared for 1970 National Incinerator Conference
    (ASME) at Cincinnati, Ohio (May, 1970).

7. Hilsheimer, H. "Experience After 20,000 Operating Hours:
    The Mannheim Incinerator," paper prepared for 1970
    National Incinerator Conference (ASME) at Cincinnati,
    Ohio (May, 1970).

8. Eberhardt, H. "European Practices in Refuse and Sewage
    Sludge Disposal by Incineration," paper prepared for
    1966 National Incinerator Conference (ASME) at New York
    (May, 1966).

9. Eberhardt, H. and W. Mayer. "Experiences with Refuse
    Incineration in Europe. Prevention of Air and Water Pol-
    lution, Operation of Refuse Incineration Plant combined
    with Steam Boilers, Design and Planning," paper prepared
    for 1968 National Incinerator Conference (ASME) at New
    York (May, 1968).

10. Stabenow, G. "Performance and Design Data for Large
    European Incinerators with Heat Recovery," paper prepared
    for 1968 National Incinerator Conference (ASME) at New
    York (May, 1968).

11. APWA Special Report. "Resource Recovery from Incinerator
    Residue," Vol. 1, *Findings and Conclusions*, APWA-SR-33
    (November, 1969).

12. Brackett, C. E. "Production and Utilization of Ash in
    the U.S.," paper presented at Second Ash Utilization
    Symposium--Ash Utilization. Bureau of Mines Information
    Circular No. 8488, U.S. Department of the Interior.

CHAPTER 11

NONCONVENTIONAL TYPES OF INCINERATION

Some of the more novel approaches to incineration of refuse include melting with auxiliary fuel (termed "total incineration" or slagging), suspension burning, and fluid bed combustion. Pyrolysis might also be included in this general category, but, strictly speaking, it is not a method of incineration; it consists, rather, in decomposing the refuse in the absence of air or oxygen by application of high heat in a totally non-oxidizing atmosphere such as nitrogen. However, certain pyrolysis systems, such as the Landgard system, may be considered as starved oxygen incinerators.

TOTAL INCINERATION

Total incineration is defined as the conversion of refuse to solidified slag and flue gases, the latter including mainly carbon dioxide, oxygen, nitrogen and water vapor. Zinn, LaMantia and Niessen[1] describe the process in great detail. The slag by-product represents the lowest possible volume of ash residue, and a number of processes are under advanced development in this area. Warner et al.,[2] Stephenson[3] and Niessen[4] bring out other aspects of these processes. The need for supplementary energy varies among these processes. All systems can be provided with adequate air pollution control.

In contrast to conventional incineration where temperatures are in the order of 980°C (1800°F), all or part of a total incineration system must operate at temperatures approaching 1650°C (3000°F) in order to convert the ash residue to a liquid slag. This slag can be quenched in water to form a granular material, or it can be allowed to cool slowly in a pit producing a solid mass, which can subsequently be broken into lavalike lumps, similar in size to crushed stone. The process is also referred to as a "slagging" incinerator. Concepts are currently in various stages of development, although they have not as yet been tested or demonstrated on municipal refuse.

The principal objectives of total incineration are:

- Maximum volume reduction of solid waste (approximately 97.5%;
- Complete combustion or oxidation of all combustible materials, producing a solidified slag which is sterile, free of putrescible matter, compact, dense and strong;
- Elimination of the necessity for a large residue-disposal operation adjacent to the incinerator; and
- Complete oxidation of the gaseous products of incineration, with discharge to the atmosphere after adequate treatment for air pollution control.

Fusion can be accomplished either by operating at temperatures above the melting point of the ash residue or by melting the ash in a separate device subsequent to conventional incineration. Temperatures in excess of 1430° to 1540°C (2600° to 2800°F) are required for fusion, with the actual temperature depending upon the composition of the ash in the refuse. However, to insure adequate fluidity of the slag, a temperature approaching 1650°C (3000°F) should be maintained in the slag zone. Some limestone, and possibly a small amount of fluorspar, is usually required to aid in fusion and in the control of fluidity; lesser amounts of flux additives are required if materials such as glass (high silica content) are removed from the refuse before being charged into the furnace. The removal of metal from the refuse feed also has a considerable effect on the flux additives required.

In theory, refuse of less than 25% moisture could be burned to produce temperatures of 1650°C (3000°F) by operation at near stoichiometric conditions, *i.e.*, 0 to 10% excess air. However, such conditions could not be maintained in a conventional furnace because of thermal damage to the grates and incinerator walls and because of the formation of clinkers.

The ability to obtain adequate slagging temperatures depends upon the following factors:[1]

- Available heating value of refuse;
- Moisture, metal and inert content of refuse;
- Level of excess air required for complete combustion; and
- Availability of supplemental energy.

Energy-balance calculations are based upon idealized and complete combustion, perfect mixing, and thermal equilibrium in the furnace. Low excess air levels or reducing conditions can be maintained in the primary chamber with combustion completed in a secondary chamber. Only the slag fusion zone must be at 1650°C (3000°F); other parts of the chamber may be at lower temperatures.

Total incineration processes under development are shown in Table 11.1. They differ in the sequence of refuse combustion, in the location and extent of the ash fusion zone, and in the method of supplying the required supplemental energy or fuel. The DRAVO/FLK process has been demonstrated on industrial refuse in Europe. The American Thermogen process has been in large-scale pilot operation for several years in the northeastern United States. The Sira and Ferro-Tech processes have been partially demonstrated on U.S. refuse with full-scale operation contemplated. The Torrax process is under design and construction supported by EPA. The electric furnace and oxygen enrichment types represent process concepts based on existing thermo-processing systems successfully used in industry. It is believed technically feasible to produce fused slag by total incineration of industrial and/or municipal refuse from any of the seven systems.

Advantages[4]

- The maximum possible refuse volume reduction (about 97.5%) is equivalent to reducing one ton of refuse with a volume of 154 cubic feet (assumed density of 350 lb/cu yd) to a residual volume of only 3.78 cubic feet (at 110 lb/cu ft).
- All putrescible matter is destroyed in producing the slag.
- The residue is suitable as a fill material.
- All odor-causing organic species will be completely consumed.
- Most of the processes utilize less combustion air than conventional furnaces. Thus, per pound of refuse burned, the weight of gas passed to the flue gas conditioning system is less for most slagging

Table 11.1

Total Incineration Processes[1]

	DRAVO/ FLK	American Thermogen	SIRA	Ferro- Tech	Torrax	Electric Furnace	Oxygen Enrichment
Added capital cost (per installed ton) over conventional incineration	None to $2000/ton depending on choice of auxiliary equipment →						
Operating cost (per ton refuse) over conventional incineration	None to about $2.00/ton (mostly energy)[a] →						
Auxiliary energy required	Some gas or oil	Coke or gas	Some gas or oil	Coke	Gas	Electric Power	Bulk Oxygen
Air preheat from recuperator	Yes	Possible	Yes	Yes	Possible	No	No
Potential NO_x air pollution	High	High	Medium	Medium	High	Lowest	High
Relative size of APC equipment	Medium	High	Medium	Medium	Medium	Low	Medium
Operating skill required	Medium to high	High	Medium to high	High	High	Medium to high	High
Shredding required	Yes	No	Yes	No	No	No	No

[a]Oxygen required is 0.3 to 0.4 ton/ton refuse, depending on moisture.

incinerators. However, in the case of slagging
incinerators, the gas cooling system must process a
larger heat load and perhaps a higher inlet volumetric flow rate, depending on temperature and excess
air levels. Part of this heat load may be removed
in a waste heat boiler or exchanged as preheat for
the combustion air, reducing the supplementary heat
requirement for the process.

   • Although some of the systems do not require
grates for support of burning refuse, other components
are exposed to high temperature fluxing and oxidation-
corrosion, which may offset the advantages from
elimination of grates.

Disadvantages[4]

   Air Pollution. The higher temperatures required
for slagging may result in substantially more nitrogen
oxide formation than is generated in normal incineration. For combustion gases at 10% excess air, the
equilibrium nitric oxide concentration rises from
200 ppm at 2000°F to 2000 ppm at 3000°F. Although
equilibrium concentrations may not be reached due
to kinetic limitations, the rate at which the
equilibrium concentration is approached increases
rapidly with temperature. The total nitrogen oxide
generation rate (lb/hr) from incineration appears
to correlate with the heat release rate (BTU/hr) as
it does in the combustion of fossil fuels. The
higher gas temperature attained in slagging incinerators would lead to expectation of higher rates
of nitrogen oxide formation. Other inorganic gaseous
pollutants derived from the combustion of refuse will
be unchanged from the amounts generated in normal
incineration operations at 1800°F. Particulate
emissions and the efficiency of necessary emission
control will vary with the design of the systems.
Those systems characterized by high gas velocities,
which tend to entrain particulate, may require more
efficient particulate removal devices to reach the
same emission level as conventional incinerators.

   Auxiliary Heat Requirement. Supplementary heat
required to maintain slagging temperatures will
result in a significant additional operating cost
above that of conventional incineration.

   Operational Complexity. The addition of ash
fusion to the incineration process complicates
operation. If slagging and incineration are performed

in the same unit, difficulties with ash fusion and
slag tapping will affect the incineration portion
of the process. If the slagging is performed in a
separate unit, the two operations must be kept somewhat in phase to prevent buildup or depletion of
residue inventory. In either type of slagging
operation, the startup process will be more difficult
than in conventional systems; this would be of particular concern if the system were shut down daily.

Safety. The high temperature operations needed
in the molten material process make slagging systems
inherently more hazardous than conventional systems.
However, handling of such materials on much larger
scales is routine in many industries.

Operational Sensitivity. Slagging processes
depend upon maintaining temperatures in a range
bounded on the low side by slag fluidity requirements
and on the high side by the limitation imposed by
the design and materials of construction. Although
the upper limit may be reasonably well defined, the
lower limit is strongly related to residue composition.
Variability in refuse heat content or residue characteristics could, therefore, demand frequent changes
in operating conditions (or feed rates of fluxing
additives). If auxiliary fuel is used, an alert
operator is necessary to minimize fuel use (operating
cost). In view of the large number of plants now
being operated under unfavorable combustion conditions
(due to operator inattention to combustion air adjustment), it is clear that greater efforts in employee
motivation or rewards may be needed to avoid high
operating costs or frequent malperformance in the
plants.

Problem Areas Inherent
in the Total Incineration[4]

Operating Labor. Because of the relatively
complex nature of these processes, greater operating
labor skills will be required than in conventional
incinerator operations. This requirement will
necessitate additional employee training, may result
in higher labor costs, and should place increased
emphasis on employee retention efforts.

Design and Materials of Construction. The high
temperatures and operational sensitivity of slagging
systems require a highly-engineered, integrated

design with special attention to the selection of
materials; a high level of process equipment and
instrumentation maintenance is necessary to ensure
acceptable unit life and reliability. The effects
of thermal cycling, if the unit is shut down daily,
may be more severe in high-temperature systems.

Slag Fluidity. Difficulties may be encountered
with poor slag fluidity (high viscosity) and solidi-
fication due to poor temperature control or variations
in the noncombustible content of the refuse. Depending upon the properties of the noncombustibles,
varying amounts of limestone or other fluxing agents
may be required.

Process Economics[4]

Capital Investment. Slagging systems contain
components which are common to conventional incineration plants:

> foundations, buildings, shops, locker rooms, and offices;
> refuse storage and handling equipment;
> furnace;
> flue-gas conditioning system;
> air pollution control system;
> residue handling equipment; and
> induced draft fan and stack.

The costs for the first two components are expected
to be the same as for conventional systems, com-
prising about 40 to 60% of the total capital investment. Slagging system furnaces will be required to
withstand higher temperatures; however, grate systems
are eliminated, except in those systems where a
slagging furnace is added after conventional incineration. The furnace portion of plant investment
for most slagging systems would thus be expected to
fall within a 20% range of conventional continuous
grate system costs. For conventional plants, this
part of the system makes up 35 to 45% of the total
capital investment.
    The air pollution control devices, flue gas
conditioning systems, fans and stack generally comprise 8 to 12% of the total capital investment of
conventional systems, depending upon the type of
control used and the flue gas volume. If the same
particulate loading and removal efficiency are
required, the savings in the cost of APC equipment
is roughly proportional to the ratio of total

volumetric flue gas rates raised to the 0.7 power
for precipitators, and to the 0.8 power for other
devices. The cost for additional gas cooling
capacity would tend to offset some of the savings.
A maximum savings of about 5% of the total capital
investment may thus be possible for slagging systems.

Residue handling equipment generally comprises
only a few percent of the total conventional plant
investment. The significantly lower ash volume would
result in savings of 50 to 75% of the cost of this
rather minor part of the total system. These savings
could be offset by the cost of slag and metal cooling
systems, particularly if slow cooling or any special
processing is attempted.

In summary, investment costs for slagging incinerator systems should not vary significantly beyond
the normal spread in costs for conventional grate
systems. Savings in residue handling and air pollution control equipment may well be offset by increased
system costs associated with the slagging aspect of
the furnace. One manufacturer gives capital cost,
including all auxiliary equipment such as air and
water pollution controls, as $12,000 per ton for a
600-ton-per-day plant.[4]

Operating Costs. Operating costs for slagging
incinerators are discussed in relation to costs for
conventional, continuous incinerators developed
earlier in this report. Particular system costs
are treated more specifically in other sections.
One manufacturer estimates operating costs at $8.00
per ton for a 600-ton-per-day plant.

Variable Costs. By far, the most significant
cost difference between slagging and conventional
incinerators is the supplementary energy input required to maintain slagging temperatures. As discussed earlier, the amount of energy required
depends upon system design, refuse composition, and
the form of the energy input. Refuse containing
28% moisture burned with 25% excess air requires a
supplementary energy input of 1.5 MM BTU/ton of
refuse burned. As an example, if energy is provided
by charging coke ($20.20/ton; 12,000 BTU/lb) with
the refuse, the slagging fuel cost is $1.25/ton
refuse (16 tons refuse/ton coke). Reduction in this
cost is possible by use of cheaper fuels such as
gas at about $0.50/MM BTU ($0.75/ton of refuse).
If energy recuperation is used to preheat the combustion air to reduce or eliminate the external
supplementary heat requirement, any operating cost

savings would be partially offset by an increase in
the capital investment required.

Electric power costs should be about the same
as for normal incinerators. For cooling of combustion gases at 25% excess air from 3000 to 600°F,
approximately 25% more water is required than is
used in conventional incinerators in cooling gases
at 100% excess air from 1800 to 600°F. At $0.30/M
gal, this is equivalent to a maximum additional
water cost of about $0.07/ton. Overall, there would
be only minor differences in utility costs for
slagging systems relative to conventional systems.

Semivariable Costs. No significant differences
are anticipated in operating labor relative to conventional systems. Any savings in residue handling
manpower would be offset by an operator tending the
slagging end of the system or handling fuel or limestone for processes involving their use. The high
temperatures of slagging systems would result in
higher refractory maintenance costs than incurred
in conventional systems. Instrumentation and control
system maintenance would also be increased. However,
elimination of grate maintenance or reduced grate
area per ton processed may compensate for the higher
refractory maintenance and result in a slightly lower
overall cost for slagging incinerators.

Fixed Costs. Since fixed costs vary directly
with the capital investment, it is expected that
these costs will be similar to those for conventional
systems.[4]

Other Economic Considerations

Metal Salvage. If the average metal content of
refuse is estimated at 8%, a maximum of 160 pounds
of metal could be salvaged per ton of refuse; however, some fraction of this would be oxidized in the
incineration process. Ferrous metal can be recovered
magnetically in the form of pellets from water-
quenched slag, or solidified after separation from
the nonmetal slag phase. The current price of scrap
iron varies regionally from $15 to $30/ton, offering
a maximum credit of $0.87 to $1.75/ton of refuse for
the slagging process. The capital investment for
such recovery equipment should be small. However,
this metal has yet to be proved valuable in the
recovered form or as feed for steel-making. For use
as a steel-making feed, the maximum copper content

of scrap is 0.3% by weight; for foundry applications,
tolerable limits are higher for some applications
with foundries representing a much smaller scrap
market than the steel industry. It is interesting
to note that zinc is not tolerable in scrap used for
steel-making because the formation of liquid eutectic
mixtures of zinc oxide with refractory constituents
tends to deteriorate the furnace refractory.

Solidified Ash. The nonmetallic phase of the
slag has potential use as an aggregate for low-
strength concrete blocks. In the tests referred to
above, in which both phases were quenched in water,
the nonmetal portion of the solid appeared to be
friable, readily crushed underfoot, and not comparable
in strength to the sand or stone used in concrete or
road building. A stronger aggregate may possibly
be produced by slower cooling rather than by quenching of the slag. Again, the commercial value of the
material has yet to be demonstrated and would vary
with local markets.

Residue Disposal Operations. The absence of
putrescible material and the decreased volume of
residue should reduce the cost and land area requirements for disposal operations. It is also likely
that land reuse may be more rapid than in the case
of untreated refuse landfills. The material may
also be used as a fill for road and construction
applications.

SUSPENSION BURNING

The suspension burning method, widely used in
power boilers, blows the finely divided fuel into a
vortex pattern in a furnace chamber so that it burns
while suspended in the turbulent air stream. It can
provide high heat release in a relatively small
volume without the necessity for supporting a burning fuel bed, grate, or hearth. This form of
combustion is discussed by Regan,[5] Pearl[6] and
Danielson.[7] Regan points out that it is well
within the limits of present technology to treat
the preparation, conveying and burning of refuse
in a manner similar to that used with coal or other
solid fuels.
There is on the market today equipment capable
of size reduction, pneumatic conveying and suspension
burning of industrial and municipal refuse. Based
on tests and experience with hogged bark and bagasse,

it was determined that the refuse should be reduced
to a maximum size of 2 x 2 inches before being conveyed either to the furnace or to a storage silo.
The reduction process consumes 15 to 20 kwh per ton
of refuse at a capital cost of $150 to $250 per ton.
While Regan and Danielson use somewhat different
terminology, they both describe a system in which,
in place of the conventional front fixed units,
single burner units are mounted in the corners of
the combustion chamber. There are usually three or
four vertically aligned burner ports per unit.
Pneumatic lines deliver the shredded refuse and the
heated combustion air to the burners which, in turn,
direct the relatively long, luminous flame tangentially
to an imaginary cylinder in the center of the furnace.
This procedure precludes the possibility of poor distribution of fuel and air. It also permits operation
with less excess air, thereby reducing the size of
the flue gas cleaning equipment. The refuse nozzles
can be tilted upward or downward to accommodate
variations in refuse characteristics and load. With
tangential firing, the fuel particles have a longer
residence time in the hottest furnace zone, thereby
assuring complete combustion of waste fuels with low
heat content. As the burning refuse particles spin
downward, additional preheated combustion air is
introduced in the lower furnace through multiple
rows of tangential nozzles. This continues the
combustion process and maintains particle momentum.
Since the larger refuse particles will not be completely burned in suspension, a small grate may be
required in the bottom of the furnace to complete
combustion of the larger particles and to remove
ash. Oil or gas firing is usually included for use
during start-up and also as a secondary fuel.

FLUID-BED INCINERATION

As mentioned by Bailie, Donner and Galli,[8] the
chemical industry has increasingly made use of the
fluidized bed in the processing of a wide variety of
fluid-solid systems. Results of this experience and
recent work on combustion in a fluidized bed have
shown that the fluid bed can provide an efficient,
compact method for the combustion of solid particles.
It is claimed that the fluidized-bed incinerator has
potential as a more economical method for the combustion and recovery of heat from waste products
involving less pollution of the atmosphere than
established methods. The work of Bailie *et al.* in

pilot plant experiments indicates that volumetric
heat generation rates in the range of 100,000 to
150,000 BTU/hr cu ft can be realized. The bed
appears stable and control of the bed simple. The
major problems encountered in the pilot plant were
associated with feeding the waste material into the
bed; however, a larger unit should alleviate such
problems.

As described by Copeland,[9] fluidized beds function by forcing a stream of gas, under carefully
controlled conditions, upward through solid particles
(such as sand) which make up the bed. The gas stream
forces a passage between the particles of the bed,
setting them in heterogeneous motion and causing
the mass to take on a fluid character. The mixing
of solid particles in a fluid is rapid and complete,
and gas-to-solid and solid-to-solid heat exchange
is excellent. Generally, it can be said that fluid
beds will burn anything which can be fed into them.
Solid wastes with a minimum of free surface water
are blown into the reactor, whereas dryer materials
can be fed by means of a sealing-type screw conveyor.
Semiplastic sludges could be fed by a progressing
cavity pump. Thermoplastic materials such as grease
are most readily fed by first being melted and then
centrifugally pumped.

In the Copeland and similar processes the
fluidized-bed zone contains the fluidized medium-
inert material such as sand or pelletized ash.
Figure 11.1 shows a sketch of the fluid-bed principles.
The fluid-bed incineration system will normally consist of the following components:

   -- the fluid-bed reactor, made up of a plenum chamber,
      an orifice plate, a fluid-bed or combustion zone
      and a disengaging or free-board zone;
   -- a main air supply, furnishing air for both fluidiza-
      tion and oxidation (combustion)
   -- a primary dust collection system
   -- a secondary dust collection or gas scrubbing system
   -- a feed and product discharge system
   -- instrumentation to make the system automatic in
      operation.

In the CPU-400 process, Bergin, Furlong and
Riley[10] attempt to develop a new incineration system
to burn municipal refuse and expand the resulting
hot gases through a gas turbine to produce mechanical
power. One of the subsystems conceived was to first
macerate the refuse to a maximum particle size diameter
of approximately 2 inches. After comminution of the

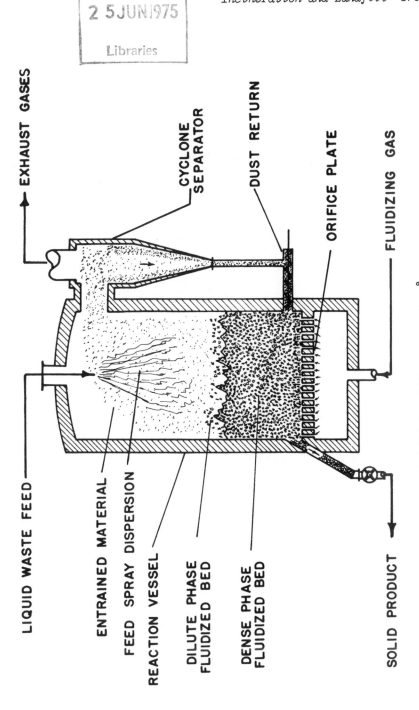

Figure 11.1. Fundamentals of fluidized solids processing.[9]

refuse to a maximum size of 2 inches, the refuse is
conveyed pneumatically to a separator to remove and
store noncombustibles such as glass, metal and rocks.
The segregated combustible portion is stored and
then passed through a dryer into the high-pressure
combustion portion of the plant. The normal considerations in fluid-bed combustors then apply.
Refuse is fed into the bed where a temperature of
815 to 900°C (1500 to 1650°F) is maintained. Combustion air at 100 psi and about 315°C (600°F) is
admitted into the bottom of the bed. Exhaust gases
pass into a cleaning system and then to a gas turbine
powering a compressor and an electrical generator.
After passage through the turbine, the still hot
gases will be used to heat the drying air or can
pass through a boiler to produce steam. Two problem
areas mentioned are: (1) development of a combustor
capable of reliable operation under high pressure,
and (2) removal of corrosive particles from the hot
gases to ensure adequate life for the turbine blades.
The next phase of this project was to investigate the
design and operation of a large-scale fluid-bed
refuse-burning unit fed by a realistic solid waste
processing system. A full-scale prototype which
will burn 400 tons of refuse per day in a fluidized
bed is planned.[10]

Advantages and Disadvantages

The advantages of fluid-bed incineration are
given by Niessen[4] as:

- Simplicity of construction. The incinerator consists of a vertical cylindrical chamber with no
  moving internal parts. High volumetric heat release
  rates can be achieved.
- Complete combustion at relatively low temperatures.
- Low NO_x emissions because of low operating temperatures and absence of local high temperature
  combustion zones or hot spots.
- Low flue gas volumetric rates.
- High heat sink capacity. Large thermal capacity
  tends to even out fluctuations in short-term
  variations in feed characteristics.
- Ease and efficiency of intermittent operation.
  Only a short reheat time is necessary prior to
  beginning incineration, even after extended
  shutdown periods.

The disadvantages are:

- Considerable preparation is needed to assure retention of particles in the bed, the complete combustion of refuse, and the removal of noncombustibles.
- Flue gas particulate loading. The high gas velocities will result in high solids loading, requiring more highly efficient particle removal equipment to achieve parity to conventional incinerators in respect to emissions.
- Operational complexity and sensitivity. Since very dense objects not supportable by the bed will drop out and interfere with fluidization, they must be removed.
- Adequate controls are needed to ensure that large increases in refuse heats of combustion do not result in extreme bed-temperature variations.
- The maximum single-unit size for refuse is estimated to be 50 to 60 tons per day, the maximum cross-section consistent with good bed distribution.
- High power consumption.
- High cost. The cost in tons per day appears to be extremely high, perhaps prohibitively so.

There do not appear to be any fluid-bed commercial installations which are arranged to incinerate household and municipal solid wastes exclusively. While the Copeland process works well on paper mill wastes and work is in progress to extend its utility to properly prepared solid wastes, including plastics, it will be some time before the process can be considered satisfactory for large industrial or municipal methods of incineration.

The best known application for fluid-bed incineration is perhaps in the handling of sewage. Dorr-Oliver equipment is well known in this field.

The CPU-400 project is currently in the category of innovation and the "proving-in" of engineering ideas and principles of fluid-bed technology. Although it is supported by federal funds, the chance of its being out of the demonstration stage and into full-scale commercial application with working installations is, at this juncture, considered remote. Figure 11.2 shows the flow diagram. No realistic costs for the method have been projected.

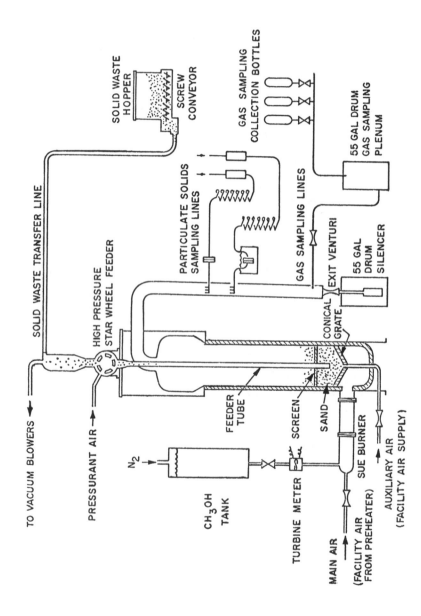

Figure 11.2.   Fluid bed combustor schematic for second series of tests.[10]

REFERENCES

1. Zinn, R. E., C. R. LaMantia, and W. R. Niessen. "Total Incineration," paper prepared for 1970 National Incinerator Conference (ASME), Cincinnati, Ohio (May 1970).
2. Warner, A. J., C. H. Parker, and B. Baum. "Solid Waste Management of Plastics," a study made for Manufacturing Chemists Association (December 1970).
3. Stephenson, J. W. "Incineration Today and Tomorrow," *Waste Age* (May 1970).
4. Niessen, W. R. *et al.* "Systems Study of Air Pollution from Municipal Incineration," Contract CPA-22-69-23, Vol. 1, PB 192378, Environmental Protection Agency, National Air Pollution Control Administration (March 1970).
5. Regan, J. W. "Generating Stem from Prepared Refuse," paper prepared for 1970 National Incinerator Conference (ASME), Cincinnati, Ohio (May 1970).
6. Pearl, D. R. "A Review of Modern Municipal Incinerator Equipment," Vol. IV, Part 4, *Technical Economic Study of Solid Waste Disposal Needs and Practices,* Public Health Service Publication No. 1886, compiled by Combustion Engineering Company for Office of Solid Wastes Management Program, Environmental Protection Agency.
7. Danielson, A. J. (editor). *Air Pollution Engineering Manual.* Public Health Service Publication No. 999-AP-40, Air Pollution Control District, County of Los Angeles, California, National Center for Air Pollution Control, Environmental Protection Agency.
8. Bailie, R. C., D. M. Donner, and A. F. Galli. "Potential Advantages of Incineration in Fluidized Beds," paper prepared for 1968 National Incinerator Conference (ASME) at New York (May 1968).
9. Copeland, G. C. "The Design and Operation of Fluidized-Bed Incinerators for Solid and Liquid Wastes," paper prepared for National Industrial Solid Waste Management Conference at Houston, Texas (March 1970).
10. Bergin, T. J., D. A. Furlong, and B. T. Riley. "A Progress Report on the CPU-400 Project," Public Health Service Publication, Office of Solid Wastes Management Program, Environmental Protection Agency (1970).

CHAPTER 12

SPECIAL PROBLEMS

PROBLEMS OF DESIGN

For incinerators in which boilers are within the system, Regan[1] points out that there are two boiler design problems that must be given special consideration when firing refuse. These are fouling of heating surfaces and potential corrosion. It is imperative that the boiler designer treat these problems correctly in order to insure a high availability factor for the refuse disposal unit.

Fouling of Heating Surfaces

In a refuse-fired boiler, all sections of the heating surface (water wall, superheater, generating bank, economizer, and air-heater tubes) are subject to fouling from a slag and fly ash deposition. Proper furnace sizing, arrangement of heating surfaces, and correct use of soot blowers can reduce fouling to an acceptable level. Proper furnace sizing means providing adequate volume and residence time to insure complete burnout. The required volume and dimensions for flame travel will vary with the method of fuel preparation, burning mechanism (*i.e.*, fixed grate, reciprocating grate, traveling grate, suspension firing, etc.), over-fire air and under-fire air system, and the amount of water cooling in the furnace. Proper furnace sizing also means providing sufficient water cooling to reduce the temperature of the products of combustion to the point where the ash is not fluid and thus will not "freeze" on boiler tubes. This type of deposit is difficult to remove with soot blowers. Also, the furnace design must provide for a temperature in the burning zone that is high enough to destroy bacteria.[1]

The next step in preventing serious slag and fly ash deposition is proper arrangement of the heating surface. This begins with adequate furnace cooling of the flue gas and is followed by using wide convection tube spacing, especially in the high-gas-temperature zones, and in-line tube rows. Staggered tube rows should not be used in a refuse-fired unit. Also, the gas velocity should be low, and retractable soot blowers should be used to clean the convection heating surface.[1]

Corrosion

When a high percentage of the boiler heat input is derived from refuse, the problem can be a serious one; inattention to it or being overly optimistic during the design stages can result in extremely high maintenance costs and greatly reduced incinerator availability. There are at least four recognized types of corrosion that must be considered by the designer. These are (1) high-temperature, liquid-phase corrosion, (2) corrosion due to nonuniform furnace atmosphere, (3) corrosion by HCl, and (4) low-temperature or dew-point corrosion. Details of corrosion problems are treated separately in the next chapter.

Other Design Problem Areas
Literature Citations

Although not, strictly speaking, primary problem areas in design, some items for future study are noted by Zinn.[2]

- Develop improved methods for collecting, feeding and handling of refuse.
- Develop improved grates and grate materials to permit the use of hot under-fire air.
- Study thermal expansion problems and develop solutions.
- Develop methods for reduction of fly ash entrainment.
- Develop improved methods for flue gas cooling.
- Develop lower cost devices for gas cleaning.
- Develop improved methods for the separation and salvage of scrap metal from ash.
- Study the utilization of the granular portions of the ash residue after scrap metal removal.
- Study the possibility of eliminating the dump site for disposal of ash residue.

• Consider the application of automatic combustion control devices for complete and trouble-free combustion with minimum labor.
• Develop improved methods for preheating and utilizing air for combustion.

For a consideration of boiler fouling and fireside tube wastage in Europe, the discourse by Eberhardt and Mayer[3] on this subject is suggested.

Design *vs.* performance specifications and performance guarantees are discussed by Stabenow.[4] While not necessarily a problem area, it is now believed possible to prepare formal codes for design and performance criteria for incinerator equipment.

Accurate incineration control is discussed by Garrett.[5] While this may not be a direct problem of incinerator design, it would appear to be distinctly related to performance. Modern incinerators do require more sophisticated control systems in order to meet increasingly stringent performance requirements. One conclusion is that the best system is not necessarily the most automated; rather, it is one that adequately meets the operational and performance requirements for the best overall initial and operating cost.

Again, although not directly related to incinerator problem areas, Kaiser[6] states that the performance of an incinerator is affected markedly by the moisture content of the refuse. He proposes that a meter be developed to monitor and report the ratio of CO_2 and water vapor in the furnace exit gas. In this work, the effects on the meter readings of firing rate, analysis of refuse, excess air, air humidity and vapor from ash quenching are shown.

PROBLEMS IN OPERATION

As pointed out by Stephenson,[7] ease, simplicity, and economy of operation and maintenance are important considerations in incinerator plant design, but the average designing engineer cannot acquire sufficient operating experience to become familiar with all of the day-to-day problems faced by plant personnel. Details considered as minor by the designer may, if not properly treated, become major problems or inconveniences in plant operation or maintenance. To bring this into sharper focus, Stephenson makes a survey of operators' views on operational and other related problem areas.

A municipal-type incinerator is one of the most difficult of process plants to satisfactorily operate and maintain. The varying composition, combustibility, and moisture content of refuse present problems in combustion control which are not encountered in plants that burn coal, oil and other relatively uniform fuels. The tendency toward wide temperature fluctuation with varying fuel, physical nature of refuse, residue, fly ash, combustion gases, and process water exacts a heavy toll in wear and tear on incinerator components and plant auxiliaries. The dirty, dusty nature of refuse and residue, and the concern over possible production of smoke, odors, or other nuisances, present continuing problems in plant housekeeping and maintenance--and in maintaining a plant's satisfactory public image. Problems in operation and maintenance vary among incinerators, including plants of similar design, and it is impossible to predict all the problems which may present themselves in a new plant. Still, ease, simplicity, and economy of operation and maintenance are and must be accepted as important considerations in incinerator plant design.

No one is more familiar with the capabilities and shortcomings of an incinerator plant than its operator. After a period of experimenting with various techniques, a competent operator can usually be expected to operate and maintain his plant at the maximum efficiency permitted by its design. Still, the perfect plant has not yet been designed or built, and whenever incinerator people meet, operators can be found in groups exchanging experiences and discussing their problems. Frequently, one operator's experience will provide the solution to another's problem, but all too frequently the problems are inherent in the design and are not susceptible to easy solution. Inasmuch as it is impossible for a designer to become familiar with all incinerator plant problems and routine procedures, the suggestions of experienced operators should be valuable as guides and "red flags" for the practical aspects of future designs. The request for such suggestions results in an enthusiastic response from leading incinerator superintendents and other municipal officials directly responsible for plant operation and maintenance.[7]

Operators generally feel that plants should be designed with firm capacity (largest unit out of service) adequate to handle the daily refuse load.

This means a spare furnace for a plant operating 24
hours a day. In plants operating less than 24 hours,
number and capacity of furnaces should be such that
with one out of service, the remaining units can
handle the load through overtime or additional shift
operation. In any event, plant capacity should be
adequate to serve the municipality for a reasonable
futurity, usually ten to twenty years, with both
population increase and possible higher per capita
rate of refuse production considered.

Based on experience in existing plants, designers
should give careful consideration to the following
specifics, as suggested by Stephenson:[7]

- Design charging hopper top dimensions larger than
  the open bucket or grapple to avoid spillage.
- Design charging hopper sides with sufficient slope
  to be self-cleaning.
- Design charging hoppers and throats to deflect refuse
  away from refractory around opening to minimize
  physical damage.
- Install charging or cut-off gates above rather than
  below the charging floor for better access, and to
  minimize spillage.
- Provide sufficient test and observation ports, but
  eliminate unnecessary doors, particularly in furnaces.
- Provide a sufficient number of undergrate observation
  ports and access doors.
- Require door handles to be a type that stays cool.
- Require tight settings throughout to minimize air
  leakage.
- Specify walls and arches that are capable of unit
  replacement and be sure they have adequate support
  and ventilation.
- Be sure there are enough refractory expansion joints,
  properly designed and located.
- Omit auxiliary burners unless they are needed for a
  particular purpose.
- Provide hydraulic operation for charging and for ash
  grates and dump grates. Operators generally seem to
  dislike pneumatic operation.
- Improve the design and construction of stoker drives
  to minimize maintenance requirements.
- Be sure that stoker drives provide sufficient adjustment to compensate for normal wear.
- Provide alarms and automatic shut-offs for stoker
  drives in case of jamming or other difficulty.
- For travelling grates, provide electrical interlocks
  to assure starting at low speed.
- Provide hydraulic, mechanical, or other nonmanual

means of removing undergrate siftings. Manual removal should not be considered except for very small furnaces.

- Install ash gates with travel parallel to trucks to minimize spillage.
- Provide manually controlled foam or water spray system for quenching siftings with other than hydraulic removal.
- Provide readily accessible cremation hearths for small animals. Some operators like facilities for log burning.
- Design flues to be self-cleaning as far as possible.
- Eliminate pockets, shelves, ledges, etc. on which fly ash may accumulate.
- Provide openings and gates in floors where chambers must be cleaned manually.
- Develop improved spray chamber designs with better resistance to the effects of moisture. Keep the amount of exposed metal in spray chambers to a minimum.
- Carefully design spray chamber flushing systems for complete cleaning of the floor. Avoid designs requiring manual cleaning.
- Specify abrasion resistant materials for ash hoppers, conveyor troughs and chains, pipes and conduits for water containing fly ash, and the like. Keep such pipes and conduits accessible for maintenance; do not bury them under floor slabs. Wherever possible, provide troughs with removable covers instead of pipes. Jet nozzles are desirable at bends in sluice troughs or pipes.
- Provide by-passes around waste heat boilers so furnaces can continue in operation when boilers are down.
- Provide protection for air pollution control equipment and induced draft fans in case of water or power failure. Suggestions include an emergency water supply and provision for dumping hot gas to atmosphere in case of fan or power failure. Both should be activated automatically in emergency.

Instrumentation and Controls Considerations[7]

Nowhere is the operators' desire for simplicity combined with ruggedness more evident than in their comments on instrumentation. They recognize that modern plants require more sophisticated instrumentation, but they generally ask that it be limited to essentials, which vary from plant to plant. Many

operators feel best operation is still achieved
by direct observation of the fire and basic manual
control, with such assistance as essential instruments and automatic controls may give.

Good intercom and public address systems are
considered essential. A PA system which is "on"
continuously is preferred to a system that turns
on only when a microphone is used.

In larger plants particularly, it is desirable
to provide indication of key functions (temperature,
smoke, etc.) in the superintendent's office as well
as in the operating area. Many operators prefer to
have all recorders in the office with only indicators
and basic controls in the operating area. Installation of duplicate thermocouples at key locations
assures an accurate reading in case one fails.
When one chimney serves two furnaces, the chimney
thermocouple should be located so as to detect the
temperature of the combined gases. Smoke detector
sleeves should be large enough to permit adjustment
of the light source and sensor.

PROBLEMS OF PERSONNEL CONCERNED WITH INCINERATOR OPERATIONS

Even with the best equipment and the most
modern and efficient procedures, all tailored for
a given job of incineration in a given town, city,
municipality, region or other locality, many
authorities in the field of refuse incineration
do not consider that completely acceptable performance can be realized unless the incineration
plant is staffed with competent managers, operators
and other employees. This includes the important
function of equipment maintenance and cleanliness.
Even such a procedure as the mixing of refuse prior
to feeding the furnace requires a certain degree
of skill, as improperly mixed refuse can cause
serious incinerator operating problems.

Comments Appearing in Literature Studies

Albert Switzer Associates and Greenleaf/Telesca[8]
in a study of refuse disposal in the tri-parish
metropolitan area of New Orleans, Louisiana comment
that: It is common practice to hold refuse in
storage pits for abnormally long periods of time,
resulting in decomposition, odor and nuisance
problems; three regularly operating incinerators

in the area could be substantially improved by better
housekeeping plus a regular program of preventive
maintenance; budgets are inadequate and unrealistic
to provide for proper maintenance, including replacement of equipment where needed, and other aspects of
preventive maintenance; personnel requirements should
be reviewed on qualifications and pay rates established
on the basis of duties and skills required.

Based on the 1968 Interim Report on the National
Solid Wastes Survey (by P. J. Black, A. J. Muhich,
A. J. Klee, H. L. Hickman, Jr., and R. D. Vaughan),
Hickman[9] states that there are approximately 300
incinerators in this country (Note: It is presumed
that this refers to municipal incinerators only.),
70% of which are without adequate air pollution
control devices. These incinerators must be studied,
evaluated and upgraded to meet the more stringent
pollution control regulations now being promulgated
in this country. In many instances, complete
abandonment of these facilities will be necessary
because of their antiquated designs and capabilities.
Because many communities are without adequate space
for sanitary landfill, new incinerator plans will be
needed. The 1968 level of the solid waste load being
treated by incineration (8%) may have to be increased
to help do the job. In the same report, Hickman
states that: The current concept of manpower required in incineration operations must be completely
changed; it can no longer be afforded to overstaff
incinerators in an effort to operate antiquated and
poorly designed facilities beyond their capabilities;
to attract and retain the manpower with the qualifications needed, salaries and benefits must be reviewed
and the solid waste management field, both public and
private, must make efforts to train personnel and
provide the opportunity for a feeling of pride in
their work.

In a summary and interpretation of the same
National Solid Wastes Survey, Vaughan[10] comments
that the estimate (1968) of the annual expenditure
to handle and dispose of the household, commercial,
municipal and industrial solid waste material in
this country is $4.5 billion per year. Even with
this impressive total expenditure, it is concluded
that present collection and disposal systems in this
country are not really adequate. Vaughan further
states that he believes 94% of existing land disposal
operations and 75% of incinerator facilities are
inadequate and that this situation must be corrected
if the environment is to be properly protected.

In discussing recommendations for research programs on air pollution reduction, Niessen *et al.*[11]
comments that municipalities should institute pilot
study groups to weigh the advantages of pooling
resources and facilities as a long-range objective
of creating an efficient, modern incineration system
for the disposal of solid waste. Better mixing
techniques must be adopted and incorporated in
incinerator systems to reduce or eliminate combustible pollutant emissions. More training of
incinerator operators should be introduced, not
just to teach them how to operate a given facility
but to orient them thoroughly in the basic technology of incineration.

Operations and Maintenance

In discussing the fundaments of operations and
maintenance, Demarco *et al.*[12] briefly cover the area
management and personnel.

As a community plans and builds an incinerator,
it should also plan for the management and personnel
necessary for its operation. The plant supervisor
should be employed several months before construction
is completed so that he can become thoroughly familiar
with each major incinerator component as installation
progresses. Operating personnel should be obtained
early enough to work closely with representatives of
the manufacturers and contractors while the incinerator is in the latter stages of construction. In
this way, the incinerator personnel can be trained
in proper operation, maintenance, and repair.

At the outset the management, including the
plant superintendent, should develop a table of
organization showing the number of shifts, number
and types of personnel per shift, and standby and
maintenance personnel. Several methods of job
classification exist; whatever method is used
should have sufficient flexibility so that incinerator personnel can be prepared for various jobs.
Rigid job titles that tend to limit operating
personnel duties should be avoided.

Staffing needs vary with the size and type of
incinerator, number of shifts, organized labor regulations (including working hours, vacations, fringe
benefits), and the extent of plant subsidiary operations, such as heat recovery and salvage. The total
man-hours required in efficient operation range from
0.5 to 0.75 per ton of solid waste processed. This

does not include man-hours for residue disposal and
major repair work.

Management must provide sufficient employment
incentives. An acceptable working environment,
equitable pay, advancement opportunities and training, retirement, employment security and other fringe
benefits are essential.

Operation guides as follows are suggested by
DeMarco *et al.*:[12]   Every plant should post a sealed
engineering drawing, pictorial flow diagram, or
scale model of the plant, showing all major components by name and function. This diagram or scale
model should illustrate how the solid waste and its
resulting gases and residues pass through the plant,
so that plant personnel and visitors may readily
understand the various components and how they
function together. At least one set of formal drawings should be maintained at the plant for reference
by operational and maintenance personnel. The local
solid waste disposal operating agency should have
copies of all engineering drawings, showing the plant
and all its components. Equipment manuals, catalogs,
and spare parts lists should be kept at the incinerator for quick reference by employees. A manual
describing the various tasks that must be performed
during a typical shift and the safety precautions
and procedures for working in various areas of the
plant should also be kept on hand.

In emphasizing the extreme importance of establishing proper operating procedures to obtain good
overall efficiency, Hall[13] considers operations as
the keynote to successful incineration. Six major
factors affecting operation are considered as well
as the importance of close cooperation between consulting engineers, manufacturers, and operators.
Hall points out that over the years much has been
learned and written about design and construction
of incinerators, furnace types, air pollution control,
refractories and a host of other data relating to
engineering design criteria, but little or no mention
of operating methods and considerations can be found
in current or past literature. The six factors considered by Hall to be essential to a successful
operating procedure are: personnel selection and
training; preventive maintenance; plant modification;
general maintenance and cleanliness; plant safety
program; and plant records keeping.

REFERENCES

1. Regan, J. W. "Generating Steam from Prepared Refuse," paper prepared for 1970 National Incinerator Conference (ASME) at Cincinnati (May, 1970).
2. Zinn, R. E. "Progress in Municipal Incineration Through Process Engineering," paper prepared for 1966 National Incinerator Conference (ASME) at New York (May, 1966).
3. Eberhardt, H. and W. Mayer. "Experiences with Refuse Incineration in Europe. Prevention of Air and Water Pollution, Operation of Refuse Incineration Plants Combined with Steam Boilers, Design and Planning," paper prepared for 1968 National Incinerator Conference (ASME) at New York (May, 1968).
4. Stabenow, G. "Performance and Design Data for Large European Incinerators with Heat Recovery," paper prepared for 1968 National Incinerator Conference (ASME) at New York (May, 1968).
5. Garrett, C. J. "Accurate Incinerator Control: An Interesting and Important Engineering Challenge," paper prepared for 1970 National Incinerator Conference (ASME) at Cincinnati (May, 1970).
6. Kaiser, E. R. "A New Incinerator Control is Needed," paper prepared for 1966 National Incinerator Conference (ASME) at New York (May, 1966).
7. Stephenson, J. W. "Incinerator Design with Operator in Mind," paper prepared for 1968 National Incinerator Conference (ASME) at New York (May, 1968).
8. Albert Switzer and Associates, Inc. and Greenleaf/Telesca. "Master Plan for Solid Waste Collection and Disposal, Tri-Parish Metropolitan Area of New Orleans, Louisiana," Office of Solid Wastes Management Program, Environmental Protection Agency, Final Report (SW-4d) on a solid waste management demonstration for the city of New Orleans under Grant No. DO1-UI-00063.
9. Hickman, H. L., Jr. "The Challenge that the National Survey Presents," in comment on *The National Solid Wastes Survey--An Interim Report* (1968) presented at the 1968 Annual Meeting of the Institute for Solid Wastes of the American Public Works Association, Miami Beach, Florida (October, 1968).
10. Vaughan, R. D. "National Solid Wastes Survey--Report Summary and Interpretation," presented at the 1968 Annual Meeting of The Institute for Solid Waste of the American Public Works Association, Miami Beach, Florida (October, 1968).
11. Niessen, W. R. *et al.* "Systems Study of Air Pollution from Municipal Incinerators," Environmental Protection Agency, National Pollution Control Administration Contract CPA-22-69-23, Vol. 1, PB 192 378 (March, 1970).

12. DeMarco, J., D. J. Keller, J. Leckman, and J. L. Newton. "Incinerator Guidelines, 1969," Public Health Service publication No. 2012, Office of Solid Wastes Management Program, Environmental Protection Agency.
13. Hall, P. B. "Operations--Keynote to Successful Incineration," paper prepared for 1970 National Incinerator Conference (ASME) at Cincinnati (May, 1970).

CHAPTER 13

CORROSION

GENERAL BACKGROUND

Corrosion of incinerators, similar to corrosion
of steam and electric generating plants, is an old
and very complex question. Much investigational
work has been carried out, and some general conclusions can be drawn from the great deal of knowledge
gathered on this subject. However, it is necessary
to examine in considerable detail not only the
materials of construction, but also the specific
operating conditions during the periods when the
corrosion is stated to have occurred before it is
reasonably safe to postulate what may have caused
any particular corrosion observed.

In any system as complex as a refuse incinerator,
with the widely varying "fuel" composition, the wide
range of temperatures encountered and the particularly
aggressive nature of the hot gases generated, it is
to be expected that the metal parts will undergo a
progressive deterioration or wastage. This factor
will have been taken into account by the design
engineer when estimating useful life and calculating
resultant costs.

A discussion on incinerator corrosion, therefore,
will deal with cases or events that seem to be excessively severe, not expected from general considerations,
or where the engineer is attempting to prolong the life
of an installation. It will be apparent, therefore,
that the greatest amount of knowledge and experience
valuable to the incinerator designer will come from
the parallel experiences of coal-fired steam generating
plants. This is especially true for incinerators
adapted for the recovery of the heat values inherent
in the disposal of solid wastes, and it is not surprising that European engineers have much knowledge
in this area.

Looking first at incinerators having heat exchangers, questions usually revolve around the actual
temperatures of operation. Thus in a plant designed
to utilize the steam for the operation of an
electricity-generating turbine, the water and steam
temperatures are considerably higher than for plants
providing steam for operating certain pieces of
equipment or for direct commercial sale.

Initially, the tube elements of the heat exchangers are protected by an iron oxide layer. When
hot flue gases containing fly ash impinge on these
tube elements, they build up deposits which may be
nonuniform in thickness, or even nonexistent in areas
in which the gas velocity is particularly high. At
elevated tube temperatures the ash may actually melt
and the condition of molten ash furnace boiler operation exists. This latter condition is quite common
for power plant installations.

Perhaps the best analysis, bearing in mind the
very complex nature of the problem pertinent to the
specific question of the role of PVC in incinerators
and its possible effect on the corrosion of certain
incinerator parts, is the published work of Fassler,
Leib and Spahn.[1] Their work on the industrial
incinerator at the Badische plant in Ludwigshafen,
Germany is particularly pertinent. The incinerator
has been in operation for several years consuming a
refuse with sufficient polyvinyl chloride and other
chlorides content to provide an atmosphere inside
of about 1.32 grains per cubic foot of hydrogen
chloride. This value compares to a near normal 0.09
grain per cubic foot in the average municipal incinerator. At the same time, the sulphur dioxide
concentration is only about half that of a municipal
incinerator. Boiler steam at 30 atmospheres pressure
and 300°C is produced, with the water wall temperatures
not exceeding 280°C and the water in the superheaters
at 350°C. The incinerator is therefore operating
below the melting point of the fly ash produced.
After 20,000 hours of operation, it was necessary
to rebuild some of the heat exchangers because of
strong corrosion between the welded areas of the
coils and spacers. Such corrosion is often due to
contact corrosion between the different materials
used in the presence of acids formed below the condensation point of the flue gases, due to an
electrolytic process. The rebuilding process
involved a complete redesign of the configuration
of the heat exchange surfaces to avoid some of the
problems observed; since then, measurements have
shown only noncritical signs of attack.[1]

Another important observation was that the fly ash deposits built up strongly on the side of the tubes facing the hot gas stream, with practically no accumulation on the other side and with a marked temperature difference from top to bottom of the row of tubes. After 3,000 to 4,000 hours the wall thickness of the tubes, where there was no fly ash deposit and the highest temperatures prevailed, had been reduced in thickness by erosion to the point where the pipe ruptured. To solve this problem, flat iron bars were welded on the back of the tubes, changing the flow pattern of the hot fly ash-containing gases. The result is that the flat iron bars are sacrifically eroded and the boiler tubes maintain their original wall thickness.

The fly ash is primarily composed of sulphites and oxides of aluminum, silicon and iron. Detailed studies of the composition of the ash at various points in the incinerator, and the respective temperatures at these points, were made. The coatings have a reddish brown color and show a layered buildup with white-grey interlayers at many of the test positions. They are brittle and quite porous. For comparison, some samples of fly ash deposits from the Dusseldorf incinerator were obtained and analyzed at the same time. The main points recorded were:

- pH values ranged from 3.1 to 5.2
- sulphate contents varied from 24.1 to 48.0%
- water solubles varied from 40.7 to 85.8
- chlorides were present in slight amounts
- no sulphides could be detected
- there were no essential differences between the Dusseldorf samples and the BASF samples
- the presence of Fe_2O_3, Fe_3O_4, ZnS and KF was definitely established
- silicates of undetermined composition were also identified.

It is interesting to note the apparent correlation between pH and sulphate content--the lower the sulphate content the higher the pH. This suggests that with a pH of 3.5 the deposits are mainly complexes of alkali iron sulphates, while at pH 2 it is principally iron sulphate. At lower than pH 2, it is alkali pyrosulphate and at pH 6 and above, alkali sulphate. It was also determined that slumping of the ash started at 470°C with complete melting by 500°C.[1]

196 Solid Waste Disposal

Because of the obviously inhomogeneous nature
of the deposits, it is not possible to accurately
analyze composition as a function of distance from
the heat exchanger wall, but it is clear that the
iron content is highest closest to the wall and the
calcium content increases towards the surface of
the deposit. The only reasonable conclusion that
can be reached is that both chloride corrosion and
alkali sulphate corrosion are present, but which
plays the dominant role can not be determined.
However, it was surmised that basically there was
no observable or assignable difference in the corrosion from that occurring in any steam-generating
plant, and that having a refuse incinerator with a
relatively high percentage of chloride present (as
hydrogen chloride principally) gave essentially no
new problems.

A most interesting observation made was that
an analysis of the deposit on the probe tube inserted
in the furnace to obtain temperature readings and
ash deposits showed a surprising difference of
chloride content with time. After only 15 minutes
in the furnace the chloride content was relatively
high and the sulphur content low compared to measurements made after six weeks. The suggestion is made
that iron chloride is first formed which subsequently
diffuses and decomposes.

CORROSION MECHANISMS

An examination of the possible corrosion mechanisms suggests that hydrogen chloride or chlorine can
diffuse through the deposits of fly ash to react
with the surface oxides and metals of the boiler
tubes with the formation of ferric chloride (300 to
400°C). The ferric chloride diffuses outward and
is converted into ferric oxides, gaseous hydrogen
chloride and chlorine at temperatures between 500
and 600°C, which in turn migrate to the surface of
the tube and keep the process going. The key
question, therefore, lies in the transport properties
of the reaction

$$Fe_2O_3 + 6 HCl \rightleftharpoons Fe_2Cl_6 \text{ (gaseous)} + 3 H_2O \text{ (gaseous)}$$

in the temperature range 300 to 400°C. The importance
of keeping the temperature of steam generation down
to avoid or minimize corrosion due to this cause is
therefore apparent.

Summarizing the possible reactions of the iron-chlorine system,[1] the process of hydrogen chloride corrosion may be expressed schematically as follows:

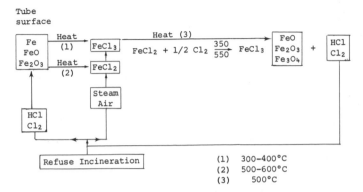

The processes shown here depend primarily on the diffusion speeds of the components in the deposits on the heat exchangers and are determined by the temperature gradients between the flue gas and the walls of the tubes. As the gas temperature is reduced, the decomposition of the $FeCl_3$ occurs further away from the surface of the tube and the corrosion attack is lessened. Under certain circumstances the decomposition will no longer take place within the coating, but rather on the border layer between the fly ash coating and the hot gas stream.

These considerations will explain why it is frequently possible to detect little or no chloride in the coatings and why only small quantities of hydrogen chloride are necessary to maintain the process. Conversely, the presence of larger amounts of hydrogen chloride in the flue gas itself does not increase the corrosion due to this mechanism. Thus, the amount of hydrogen chloride-generating materials in refuse in the absence of polyvinyl chloride itself is sufficient to initiate this series of reactions, and doubling or tripling the amount of polyvinyl chloride in the refuse will not, by itself, cause any more corrosion.

Nelson and Cain[2] describe a corrosion mechanism and show that the formation of alkali ferric sulphates is responsible for the corrosion damage. This interpretation can explain much corrosion damage, although it cannot be accepted without reserve. While related to coal dust fired burners, the results can be applied, with reserve, to the conditions in an incinerator since a complete dissociation of the

sodium and potassium salts occurs whether or not the
salts are present in the coal or in the combustion
products other than sodium sulphate, sodium chloride
and other volatile sodium salts.

The causes of the corrosion are the compounds
formed by reaction between sulphur (or its combustion
products SO_2 or SO_3), the oxides of sodium or potassium found in the ash, and the iron oxides normally
present on the surfaces of the boiler tubes. Sodium
and potassium oxides can react with the SO_3 present
in flue gases to form sodium and potassium sulphates.
At temperatures between 320° and 480°C, alkali
pyrosulphates ($K_2S_2O_7$ or $Na_2S_2O_7$) can develop which
will further react with ferric oxide to give the
complex alkali ferric sulphates:[3]

$$Fe_2O_3 + \underbrace{3\ K_2SO_4 + 3\ SO_3}_{3K_2S_2O_7} \xrightarrow[480]{320} 2\ K_3Fe\ (SO_4)_3$$

At still higher temperatures, 550° to 700°C, these
alkali ferric sulphates can become molten, at which
time they are extremely aggressive and lead to
severe corrosion attack. A considerable amount of
oxygen is consumed in the corrosion due to the
formation of the alkali ferric sulphates, and since
the gas exchange with the flue gas is rendered difficult close to the metal because of the small size
of the pores in the coating, the oxygen in the
pores is depleted and the CO_2 from the flue gases
is reduced to CO. The resultant CO can then cause
carburization. The observation that carburization
is most intense where the corrosion is most severe
supports this hypothesis and leads to the generally
accepted thesis that to minimize corrosion in
incinerators, combustion must be performed always
in an oxidizing atmosphere.

The conclusions of Fassler *et al.*[1] are
informative:

- The severity of the attack is determined primarily
  by the temperature of the tube wall and the
  temperature gradient between flue gas and tube wall.
- An electrochemical (acid) corrosion attack occurs
  below the condensation point of the flue gas.
- High temperature corrosion begins above a critical
  temperature, which in this case is greater than
  320°C.

- Three types of attack can be recognized: (1) corrosion in the gaseous phase; (2) corrosion due to ferric chloride or alkali ferric sulphate formation; and (3) corrosion due to ferric chloride or alkali ferric sulphate decomposition.
- The attack in the gaseous phase is much less than in the melt.

In a paper by Rasch[4] dealing specifically with the role of sodium chloride in high temperature corrosion, reference is made to the BASF work. The fact that any sodium chloride present in the refuse can have a sufficiently high vapor pressure at furnace temperatures to take part in corrosion reactions is brought out, but otherwise the paper supports the general data discussed earlier.

It can, therefore, again be stated that to avoid corrosion in incinerators employing heat exchangers for waste heat recovery, it is important to operate at temperatures above the point at which condensation can occur, below the temperature at which melting of the fly ash takes place, and in a strongly oxidizing atmosphere. A suitable temperature range in the heat exchangers would be between 150 and 450°C, and preferably 150 and 350°C.

In considering the question of corrosion of incinerator parts where recovery of heat values is not practiced, it is clear that any materials of construction attacked by hydrochloric acid (that is, where the temperature is below the dew point, or water is present to dissolve the hydrogen chloride) or by other acids present (sulphuric and nitric, or organic acids such as acetic) should be avoided. Since incinerators are usually designed to operate continuously, the danger of corrosion in the main furnace is very minimal (especially as the lining is principally fire brick) and occurs chiefly during periods of shut down if sufficient hot air has not been previously passed through the incinerator to sweep out the acid-containing gases. It is also clear that acid resistant parts should be provided for all spray nozzle and pipe systems used in water cooling of the flue gas and for fan blades, housings and parts where acid-containing gases might condense. In a Venturi scrubber system for removal of particulates, acid resistant materials should also be specified. With the modern knowledge available, there is no reason for permitting continuation of practices that give rise to some of the corrosion situations that have been cited.

TYPES OF INCINERATOR CORROSION

There are at least four recognized types of corrosion in incinerators.[8,9] These are (1) high-temperature, liquid-phase corrosion, (2) corrosion due to nonuniform furnace atmosphere, (3) corrosion by acid, and (4) low-temperature or dew-point corrosion.

High-Temperature, Liquid-Phase Corrosion

This form of corrosion is probably caused by molten alkali-metal sulfates. Much has been written about this type of corrosion, and some disagreement seems to exist among investigators on the exact corrosion mechanism and the temperature range in which it occurs. However, it is generally agreed that this type of corrosion is likely to occur at metal temperatures above 900°F. Recent data indicate that high gas temperatures aggravate the problem.

Corrosion Due to Nonuniform Furnace Temperature

This type of corrosion is caused by the products of partial combustion in reducing atmosphere. A reducing atmosphere can be present locally in a unit supplied with 200% or more excess air as a result of stratification or improper air or fuel distribution. Thus, carbon monoxide or hydrogen sulfide can be produced. It is thought that these compounds attack and cause failure of water-wall tubes by reducing the iron oxide on the surface of the tubes. It is important, therefore, that the burning system provide not only the correct fuel-air ratio but also proper distribution of air to the fuel and sufficient turbulence to prevent stratification.

Corrosion by Acid

The third type of corrosion, resulting from acid, has been recognized for years. Sulfur oxides and hydrogen chloride (HCl) are introduced from the combustion of paper, fuel oil, and organic matter; in addition, HCl comes from salt and polyvinyl chloride.

Low-Temperature or Dew-Point Corrosion

Finally, low-temperature or dew-point corrosion occurs when the flue gas contacts surfaces that are at temperatures below the dew point of corrosive constituents in the gas. Temperatures low enough to cause acid condensation may be found on the water-inlet end of an economizer if the feed water temperature is too low. In this instance, the economizer performs as a feed water heater, receiving water from the boiler feed pumps and delivering it, at higher temperature, to the steam-generating equipment. Other areas of low temperature that permit acid condensations are in the cold end of an air heater and on outer casings of exchanger equipment. For this reason, welded-wall construction is considered desirable. The tubes and fins are welded to form a self-cased, pressure-tight envelope to prevent flue gas from contacting a cold casing. Low temperature corrosion could also be a problem during unit outages. Some deposits are corrosive; where the deposits are hygroscopic, the problem becomes more severe as the length of the outage increases. If a lengthy shutdown is contemplated, the fire side of the unit should be water washed or, alternatively, the unit may be kept hot by using an external source of heat.[9]

FIELD OBSERVATIONS ON CORROSION

Miller and Krause[5] discuss field work involving operation of corrosion probes and/or analyses of gases and deposits in municipal incinerators in three U.S. locations--(a) Miami County, Ohio; (b) Oceanside, New York; and (c) the Navy Salvage Fuel Boiler at Norfolk, Virginia. Examinations of the surfaces of these probes and resultant deposits using X-ray diffraction and the electron microprobe have shown that compounds containing sulfur, chlorine, lead, zinc and potassium are contributing to the corrosion reactions.
Results showed that the corrosion of the carbon and stainless steels used increases as the metal temperature rises from about 350 to 1000°F. The temperature gradient moving from the metal through the deposit to the surrounding gases is also of importance.
These field studies also included analyses of furnace gases from several incinerators. Average

concentrations of corrosive gases such as SO_2 and
HCl were found to be about 100 ppm. Laboratory work
and field results provided an explanation of the
corrosion reactions. As might be expected, they are
complex and interrelated. It is suggested that HCl,
Cl_2, $K_2S_2O_7$, $KHSO_4$, NaCl, $ZnCl_2$, and $PbCl_2$ are involved. At 600°F, the most corrosive salts in
decreasing order of activity were: $KHSO_4$, $K_2S_2O_7$ and
$ZnCl_2$. $PbCl_2$ was also very corrosive.

In further discussion, Miller and Krause state
that the corrosion probe results agree quite well
with those observed in practice. For example,
Fassler *et al.*[1] suggest that high temperature corrosion begins near 600°F. Hilsheimer[6] reported
accelerated corrosion in high temperature areas of
water-wall incinerators in Germany with tube life
of less than a year occasionally seen.

Comparison of these indications with experience
in U.S. power plants is difficult because so few
data of this type are available. Typically, it is
believed that corrosion rates in superheaters and
reheaters will average about 0.020 inch per year.
Correlation of the results of the different studies
made on the deposits and corroded surfaces combined
with the laboratory results provided at least a
possible explanation of the corrosion mechanism.[5]

It is proposed that HCl and Cl_2 released adjacent to the tube surfaces are important factors.
In addition, SO_2 and SO_3 gases, along with sulfur-
containing compounds, cause additional corrosion.
The roles played by the sulfur- and chlorine-
containing compounds in the refuse are of great
importance and are closely interrelated. The sequence
of chemical reactions that are involved in the corrosion mechanism is depicted in Figure 13.1.
Chlorides and oxides reach the tube surface by
direct volatilization in the flames and by reaction
of the HCl formed during burning with the K_2O and
Na_2O volatilized. Sodium salts are shown in Figure
13.1 by way of example, but similar reactions occur
with potassium salts. Chloride salts deposited on
the metal surface react with SO_2 and oxygen near the
tube to evolve high concentrations of HCl directly
adjacent to the metal. Some of this HCl reacts
directly with the iron to form $FeCl_2$. However, a
more serious condition may involve catalytic oxidation of the HCl to Cl_2. The Cl_2 is much more
reactive with the tube metal and can take part in
a closed cycle of reactions in which the product
$FeCl_2$ is converted to Fe_2O_3, and Cl_2 is regenerated
as shown on the left side of the diagram.

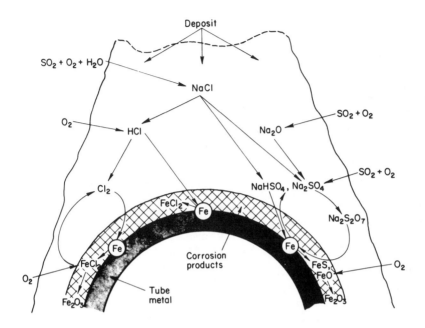

Figure 13.1.   Sequence of chemical reactions explaining corrosion on steel tube.[5]

     The additional role played by sulfur is that of forming low-melting pyrosulfates or bisulfates by reaction of sulfates in the deposit with additional sulfur oxides. This action is shown on the right side of the diagram, where another closed loop is possible. In this case, pyrosulfates or bisulfates react with the iron to form FeS and FeO and to regenerate Na_2SO_4. The corrosion process is further complicated by the presence of zinc and lead salts which serve to lower the melting points of the mixtures on the metal surface.

     It is concluded that water-wall incinerators using the metals evaluated should not be designed to generate superheated steam, but should be operated at relatively low metal temperatures, near 500°F, to minimize corrosion. The down time of an incinerator can be important from the corrosion standpoint since acidic salts can become wet by absorption of water when the surfaces become cool.

     Thoemen[7] discusses the experience with tube corrosion during 6.5 years of operation at the Düsseldorf, Germany incinerator plant. This plant has passed an operating time of more than 43,000 hours for each of its four units and has consumed

a total of about 1.6 million metric tons of solid
waste. In covering the history of tube corrosion,
Thoemen details the location of corrosion progress
in his study. In 1966 a large plant reported tube
corrosion in the first boiler pass behind the incinerator furnace. Subsequently in other large plants
the same phenomenon appeared with temporal displace-
ment according to the operating conditions and times.
The experiences accumulated at Düsseldorf may be
used as an example.

For the first time after about 1000 hours of
operation time, comparatively severe corrosion was
experienced on the tubes of the first stage superheater at the side of the direction of gas flow.
The uppermost rows of tubes of the superheater
carried, under a relatively small scale of deposits,
a heavy layer of corrosion products. By the thickness of this layer, it had to be concluded that
rapid wastage of these tubes could be expected.
However, as operations continued, the corrosion
rate declined and has reached a level which causes
a barely measurable waste of material. The tubes
now are covered with a hard layer of deposits.

In 1967 after more tube failures had been reported at other incinerator plants (in Düsseldorf
the tubes just above the furnace were inspected and
at the lower area of the side walls), the first
indications of corrosion were found. The appearance
of corrosion products pointed to destruction of the
material by reducing gas atmosphere in combination
with hydrochloric acid. As a means of protection,
additional secondary air nozzles were installed to
build up a curtain of excess air in front of the
tubes. Furthermore, the endangered area was studded
and covered with a 1/2-inch layer of silicon carbide
refractory. In other plants this method of protection
already has been applied with similar good results.

In 1968, a tube rupture occurred in one of the
boilers. The cause was investigated and found to be
corrosion by the flue gas. Subsequent extensive
inspections of the boilers demonstrated that corrosion
observed the year before had continued and reached
the tube surface above the protected area. Not only
the side walls but the front wall was affected.
Ultrasonic measurements showed that it was necessary
to renew about 30 to 40 tubes in each boiler for a
length of about 7 feet. A considerable number of
the remaining tubes had to be reinforced by welding.
Moreover, an extended area of tube surface was
studded and concealed.

In 1970 a tube failure was experienced on the final stage superheater which hangs in the upper part of the first boiler pass. The side of the tubes facing the gas flow showed severe corrosion. At first, the same causes were presumed as found in the lower part of the gas pass. Samples and tests of the flue gas proved, however, that in this case another mechanism was involved--*erosion*. To protect the tubes against the gas flow, flat steel bars were fixed to the affected side of the tubes. This preventive measure has had good results for 1-1/2 years; however, tests on tubes of a material which shows promise of withstanding the attack are in progress.

Thoemen postulates that the waste of tube material at different parts of the boiler can be based on three different reactions:

(1) In the first boiler pass corrosion occurs due to the absence of O_2 in the flue gas. In this case, $FeCl_2$, which evaporates at operating temperatures, is formed on the tube surface.

(2) The corrosion on the superheater tubes is caused by chlorine ions or chlorine in *statu nascendi* set free by the formation of sulphates from condensing alkali compounds:

$$2K \ Cl_2 + SO_3 + 1/2 \ O_8 K_2 SO_4 + 2 \ Cl$$

(3) Other observed tube defects especially in the second or third pass are mainly caused by a combination of erosion and corrosion.

This proven knowledge from the chemical side fits the conclusions reached by looking at the combustion technique from a practical point of view. The areas with low excess air coincide with the corroded tube surfaces, whereas the protected area becomes part of the furnace without the danger of corrosion.

THE ROLE OF PLASTICS

With plastics currently contributing approximately 2% to collected solid waste in the United States[8] and projected to increase to 4% by 1980, there is growing concern regarding their effect on the performance of commercial incinerators. Of the major commercial plastics, the one causing the most concern is polyvinyl chloride (PVC) because on burning it liberates HCl, which similar to the

H_2SO_4 from SO_2 formed during incineration, can
contribute to corrosion under improper operating
conditions.

Polyvinyl chloride is the second largest volume
plastic in the United States and the most versatile
in its variety of formulations, properties and uses.
Thus, it is important to examine and resolve any
problems which might result from the disposal of an
increasing amount of this material by incineration.

European experience has shown that modern
incinerators operating on modern principles are
entirely capable of handling considerably larger
quantities of polyvinyl chloride than presently
contained in solid wastes with no deleterious effect
on either the incinerator or the atmosphere. Some
of the U.S. incinerators, however, are old and may
not always be operated according to modern principles.
If corrosion does occur, incinerators with heat exchange facilities would have corrosion problems on
the fire side of exchange equipment when the combustion gases contact the outer metal surfaces of this
equipment. The heat exchangers will also experience
tube wastage in the steam-carrying sections on the
fire side of the internal walls. Outside of the
heat exchange section, the reheat and low stack
areas sometimes present problems. Incinerators
without heat exchangers would have their major
corrosion problems on metal surfaces above (and
sometimes below) the refractory in combustion
chambers, in the cooling area before gas cleaning,
in the gas cleaning operation, and sometimes in the
reheater before discharge to stack. As discussed
by Regan,[9] there are two heat exchanger design
problems that must be given special attention when
firing refuse. These problems are (1) fouling of
heating surfaces and (2) potential corrosion. It
is imperative that the heat exchanger designer
treat these problems correctly in order to insure
a high availability factor for the refuse disposal
unit.

Table 13.1 shows general information on the
combustion products of the principal plastics products occurring in municipal wastes. Plastics
constituted 2% of all collected municipal waste in
1969. Polyvinyl chloride (PVC) constitutes about
10% of the plastics in municipal refuse, or about
0.2% of the total collected municipal wastes.
Hydrogen chloride in the combustion products from
PVC is the most significant corrosion-producing
acidic component found in the plastics portion of

municipal waste.[10]   Outside of the miniscule amounts
of hydrofluoric acid from the very low concentration
of fluorine polymer in waste, none of the other
plastics shown in Table 13.1 contribute acidic
combustion products.

Table 13.1

Plastics Decomposition Products[8]

Plastics	Examples of Applications	Combustion Products Other than Water and Carbon Dioxide
Polyethylene and related plastics	Packaging materials, pipes, cable insulation	Carbon monoxide
Polystyrene	Packaging materials, household goods, toys, refrigerators	Carbon monoxide, carbon black
Polyvinyl chloride	Pipes, film, flooring materials, furniture, electrical insulation, packaging materials	Carbon monoxide, hydrogen chloride
Fluorine polymers	Special hospital equipment, corrosion-resistant or anti-friction coatings	Hydrofluoric acid, fluorinated hydrocarbons, carbonyl fluoride
Nylon	Fibers, gearwheels, household goods	Carbon monoxide, nitrogen-containing gases
Polyesters	Plastic boats, lacquers, building materials, fibers	Carbon monoxide
Melamine resin	Adhesives, lacquers, household goods	Carbon monoxide, nitrogen-containing gases
Phenolic molding compositions	Adhesives, handles, telephones	Carbon monoxide, carbon black

The work of Eberhardt and Mayer[11] contains a
section on emissions of sulfur oxides and hydrogen
chloride from refuse incinerators in West Germany.
Table 13.2 shows the data.

Table 13.2

*SO_2 and HCl Emissions
of Refuse Incinerators*[11]

	SO_2 Grains/ft³	HCl Grains/ft³
Municipal incinerator plant, approx. 30% industrial waste	0.122–0.163	0.122–0.204
Incinerator for industrial refuse, 50% chem. waste	0.122–0.286	0.448–1.182
Oil-fired boilers	0.815–1.020	

One grain   = 0.065 gram
One cubic foot = 0.0282 cubic meter

Other comments are as follows:

In the case of refuse incineration, 15 to 20% of the
sulfur oxides are present as SO_3, while the SO_3
fraction amounts only to 1 to 2% in oil-burning
furnaces. The sulfur content of heavy fuel oil
amounts to between 0.5 and 1.0% (in Germany), while
it is 0.3% in refuse. The higher concentration of
sulfur oxides in oil-fired boilers can be explained
by low excess of air and the smaller volume of flue
gas compared to refuse boilers in which the process
takes place with the tenfold volume of flue gas.

In these particular tests, free chlorine could not be
found in the flue gases even with extremely high
chlorine percentage in the refuse--it was always in
the form of hydrogen chloride. It is of interest that,
although the SO_2 concentration amounts only to 10 to
20% of the value in oil-fired steam boilers, the SO_3
content is higher than in oil-fired systems.

At the present time[10,12,13] the amount of polyvinyl chloride compounds in the solid waste stream
going to incineration is probably averaging about
0.2% and consists in large measure of plastic bottles,
tubes and containers, some film, and a smaller proportion of other articles such as floor tile, garden
hose, plastic raincoats, scrap wire and cable. The
actual chlorine content is lower than at first expected.
Thus, while the pure resin has a chlorine content of
56.8%, a typical bottle might only contain 49% chlorine,
and a flexible raincoat or a wire and cable compound
might be as low as 34%. This is well illustrated in
the Kaiser and Carotti report[10] where an analysis was
made of the actual polyvinyl chloride plastics used
in the experiments performed. Representative samples
of the plastics mix of bottles, film and molding
scrap were chopped in a Wiley mill to pass 2-mm
round openings and were analyzed to give:

Moisture	0.20%
Carbon	45.04
Hydrogen	5.60
Oxygen	1.56
Nitrogen	0.08
Sulfur	0.14
Chlorine	45.32
Ash	2.06
	100.00%

The chlorine content of 45.32% together with an
oxygen content of 1.56% bears out the above discussion
as well as suggests that some of the bottles were
actually a copolymer resin.

Kaiser and Carotti give a total chlorides figure
of 0.50%, while recognizing that less than half comes
from any plastics source. In a very comprehensive
study made in Hamburg, Germany, by Reimer and Rossi[14]
and reported in 1970, it was found that, on the same
basis of a chlorine content of the polyvinyl chloride
plastic as for Kaiser and Carotti, namely 45%, the
average chloride content attributable to the plastics
was 54.7%, with a range from a low of 38.1% to a
high of 74.4%. This is in good accord with the U.S.
data cited.[8] Swedish authorities also recognize
that appreciable amounts of chlorine compounds are
already present in the refuse, even in the complete
absence of plastics, as do investigators in the
United Kingdom, Switzerland and other countries.[13,15,16]

After an initial drying, when steam is evolved,
there is a period in which volatile organic matter

is evolved due to various processes of devolatiliza-
tion, degradation and depolymerization. This may
start at temperatures just above 100°C and continue
up to about 300°C. During the latter part of this
period, ignition commences. In the case of PVC, in
the range of about 170°C to 300°C, the material decomposes to give off hydrogen chloride. At temperatures
above 300°C, ignition is well advanced and the main
burning and combustion occurs from 400°C to 700°C
when the gaseous products are a mixture of water,
carbon monoxide and carbon dioxide together with the
HCl from any chlorine-containing compounds present.
During this period, any inorganic chlorides present,
such as salt, will evolve HCl.

It has been erroneously suggested that phosgene
($COCl_2$) is evolved in the burning of PVC. Well
documented studies, particularly by E. Boettner *et
al.*[17] show no generation of phosgene in pyrolysis of
PVC.

Mechanism of Corrosion During
Incineration of Polyvinyl Chloride

Corrosion by hydrogen chloride was known to
occur in power plants long before the advent of
polyvinyl chloride. The chloride ions present in
fuels as an impurity were known to be the source of
the HCl. In refuse incinerators, salt, paper and
other organic matter may contribute an amount of HCl
perhaps equal to or greater than that arising from
the burning of PVC.

Experiments in Sweden and in Germany indicate
that under oxidizing conditions the oxides normally
appearing on steel are not attacked by dry HCl.
Only when the surface oxides are reduced, leaving
the surface vulnerable, can the underlying iron of
the heat exchanger react to form a volatile ferrous
chloride. Oversimplified, the reaction is approximately as follows:

$$Fe_2O_3 + H_2 \text{ (reduction} \longrightarrow FeO + H_2O$$

$$FeO + 2 HCl \longrightarrow FeCl_2 + H_2O$$

Another factor, somewhat neglected to date in
incinerator corrosion, is the role of aluminum in
the reduction of iron oxide. The aluminum is present
in the refuse from an increased volume of aluminum
cans, a phenomenon almost unique to the United States.

$$Al + Fe_2O_3 \longrightarrow Al_2O_3 + Fe$$

Aluminum can also react with PVC, forming aluminum chloride ($AlCl_3$) which is highly corrosive--especially when molten--and which is a potent catalyst for a number of corrosion reactions. When PVC decomposes during incineration, the chlorine evolved occurs in the flue gases as hydrogen chloride; salt decomposition evolves chlorine.

Rasch[15] of the Battelle Institute in Frankfurt, Germany describes the equilibrium between chlorine, air, and HCl with the following equation:

$$2HCl + 1/2\ O_2 \rightleftharpoons H_2O + Cl_2 \qquad (1)$$

His studies reveal that the content of elemental chlorine is higher at low temperatures. Under furnace conditions, *i.e.*, above 800°C, only hydrogen chloride is present in the flue gas. The proportion of HCl in the flue gas rises with increasing content of water vapor and decreases with increasing oxygen content.

High temperature corrosion, particularly the external tube corrosion occurring in heat exchangers of refuse incinerators, is initiated by the decomposition of the oxide film protecting the metal. The oxide film is decomposed either by reduction or because alkali pyrosulfates decompose it with formation of complex alkali-iron sulfates. After the decomposition of the protective oxide film, hydrogen chloride contained in the furnace gases may react with ferrous oxide, ferric carbide and elemental iron to give volatile ferrous chlorides. Iron carbide and elemental iron occur only as unstable intermediate phases. The following reactions take place between these iron compounds and hydrogen chloride:[15]

$$Fe_2O_3 + 4\ HCl = 2\ FeCl_2 + 2\ H_2O + 1/2\ O_2$$

$$Fe_3O_4 + 6\ HCl = 3\ FeCl_2 + 3\ H_2O + 1/2\ O_2$$

$$FeO + 2\ HCl = FeCl_2 + H_2O$$

$$Fe_3C + 6\ HCl = 3\ FeCl_2 + C + 3\ H_2$$

$$Fe + 2\ HCl = FeCl_2 + H_2$$

Rasch[15] also points out that an accessory in fossil fuels is sodium chloride. At elevated temperatures sodium chloride reacts with sulfur trioxide or with silicic acid and water vapor to form sodium sulfate or sodium silicate. This reaction yields hydrogen chloride, one of the corrosive components in flue gas.

The equilibrium possible between ferric and ferrous chloride is:

$$2\ FeCl_3 \rightleftharpoons 2\ FeCl_2 + Cl_2. \qquad (2)$$

Since the chlorine component occurs in furnaces principally as hydrogen chloride, equations (1) and (2) can be combined to:

$$2\ FeCl_3 + H_2O \rightleftharpoons 2\ FeCl_2 + 2\ HCl + 1/2\ O_2.$$

Increasing temperature favors the equilibrium to ferrous chloride.

Ferrous oxide completely reacts with hydrogen chloride to give ferrous chloride. At higher temperatures conversion declines and at 900°C is less than 2%. Thus, corrosion decreases with increasing temperature until scaling occurs. Rasch indicates that his calculations and laboratory experiments prove that the reaction between iron oxides and hydrogen chloride requires a reducing atmosphere.

Wogrolly[18] studied the behavior of PVC-containing refuse in refuse combustion plants. Using a rotary furnace refuse disposal plant as an example, Wogrolly measured the relation between the plastics content of household refuse and the composition of the resultant waste gas. These tests showed that, provided incineration is carried out properly, household refuse such as is generally encountered in Central Europe today has no harmful effect on the combustion process, has no corrosive effect on the incineration plant, and creates no atmospheric pollution problems, even if the plastics content is well above average (up to 10% polyethylene and 5% PVC).

PREVENTION OF CORROSION

Recent Swedish[19] experiments show that during incineration only half of the theoretical quantity of HCl from the known amounts of chloride present can be found in the flue gases. During combustion of refuse the ash becomes alkaline and the metal oxides of the ash neutralize at least part of the HCl. A definite correlation was also observed between the temperature of the flue gas and the ability of the fly ash to absorb and neutralize hydrogen chloride.

The addition of lime to the refuse is said to aid in binding the HCl. This, however, involves the formation of calcium chloride, which reportedly would

almost entirely decompose during combustion in the
presence of water vapor to give HCl. If the lime
is mixed in the flue gases, an even greater portion
of the HCl should be neutralized without the
hydrolytic decomposition of the calcium chloride.

The recent Dow U.S. Patent 3,556,024 claims
that alkalies added to the wastes before burning
neutralize more than 75% of the HCl which would
otherwise cause corrosion and be released to the
atmosphere during the incineration of chlorine-
containing plastics.

In summary of corrosion from plastics and other
acid releasing materials, acid corrosion may occur
in refuse incineration plants under certain conditions,
partly in the steam pipes of the heat exchangers,
partly on the air cleaning equipment for separating
dust from the flue gases, and on the outermost end
surfaces of tube banks and combustion chambers.
Both theoretical calculations and practical experience
show that the oxides normally occurring on steel
(Fe_2O_3 and Fe_3O_4) are not attacked by acidic gases.
Corrosion of steel by acid does not, therefore, occur
under oxidizing conditions (an excess of air). It
only occurs under reducing conditions (insufficient
air), which may exist both locally and periodically
where the oxide surfaces are reduced and the acid
reacts with the steel to produce readily volatile
iron chlorides which do not form a protective coating.
The reduction of the surface oxides takes place very
slowly below 300°C, however, and perceptible corrosion does not, therefore, occur below this temperature.
On the other hand, the temperature in the purification
plant and the flue gas system must not be allowed to
fall below 150°C, since this would cause dew point
corrosion, *i.e.*, condensation of acidic gases leading
to corrosion. Corrosion of steel by H_2SO_4 is caused
by compounds formed by reaction between sulfur
oxides, (*e.g.*, SO_2, SO_3), the oxides of sodium and
potassium found in the ash, and the iron oxides
normally present on the surfaces of the heat exchanger tubes. At temperatures between 320 and 480°C,
alkali pyrosulfates ($K_2S_2O_7$ or $Na_2S_2O_7$) develop which
can further react with ferric oxides to give complex
alkali ferric sulfates, *e.g.*, $K_3Fe(SO_4)_3$. At still
higher temperatures (550 to 700°C) these alkali
ferric sulfates become molten and extremely aggressive,
leading to severe corrosion attack.

Some basic considerations for proper incinerator
design and operation to prevent corrosion are:

- Homogenization of refuse
- Operation always under oxidizing conditions, *i.e.*, with a substantial excess of air
- Turbulent flue gases
- Parallel-flow incineration
- Temperature on the heat exchanger between 150 and 300°C
- Temperature in the furnace proper above the dew point and below 1000°C

It can be safely asserted that present day incinerators of reasonable design and operation can handle domestic, municipal or industrial wastes containing upwards of ten times the present levels of PVC content without giving rise to corrosion or to gas emissions having hydrogen chloride levels in excess of reasonable safety limits.

REFERENCES

1. Fassler, K., H. Leib, and H. Spahn. *Mitteilungen der VGB, 48(2)*, 130 (1968) English translation.
2. Nelson, W. and C. Cain. *Trans ASME Series A, 82*, 194 (1960); and *83*, 468 (1960).
3. Corey, R. C., B. J. Cross, and W. T. Reid. *Trans ASME 67*, 289 (1965).
4. Rasch, R. *Energie, 23(2)*, (February, 1971).
5. Miller, P. D. and H. H. Krause. "Corrosion of Carbon and Stainless Steels in Flue Gases from Municipal Incinerators," paper prepared for 1972 National Incinerator Conference (ASME) at New York (June, 1972).
6. Hilsheimer, H. "Experience After 20,000 Operating Hours-- The Mannheim Incinerator," paper prepared for 1970 National Incinerator Conference (ASME) at Cincinnati (May, 1970).
7. Thoemen, K. H. "Contribution to the Control of Corrosion Problems on Incinerators with Water-Wall Steam Generators," paper prepared for 1972 National Incinerator Conference (ASME) at New York (June, 1972).
8. Baum, B. and C. H. Parker. "Incinerator Corrosion in the Presence of Polyvinyl Chloride and Other Acid Releasing Constituents," paper prepared for Society of Plastics Engineers (SPE), Regional Technical Conference (RETEC) at Chicago (October, 1972).
9. Regan, J. W. "Generating Steam from Prepared Refuse," paper prepared for 1970 National Incinerator Conference (ASME) at Cincinnati (May, 1970).
10. Kaiser, E. R. and A. A. Carotti. "Municipal Incineration of Refuse with 2% and 4% Additions of Four Plastics,"

report of work sponsored by Society of the Plastics Industry, 1971; also in Proceedings 1972 National Incinerator Conference (ASME) at New York (June, 1972).

11. Eberhardt, H. and W. Mayer. "Experiences with Refuse Incineration in Europe--Operation of Refuse Incineration Plants Combined with Steam Boilers, Design and Planning," paper prepared for 1968 National Incinerator Conference (ASME) at New York (May, 1968).

12. Warner, A. J., C. H. Parker, and B. Baum. "Solid Waste Management of Plastics," study for Manufacturing Chemists Association (December, 1970).

13. "Plastics from an Environmental Standpoint," report from the Plastics Environment Committee of the Royal Academy of Engineering Sciences, Sweden.

14. Reimer, H. and T. Rossi. *Müll und Abfalle.* (March, 1970), pp. 71-74.

15. Rasch, R. *Battelle Information, 5,* 18 (August, 1969), Frankfurt, Germany (English translation).

16. Ranby, B. "Plastics from an Environmental Protection Point of View," paper presented at EFTA Plastics Association meeting at Oslo, Norway (May, 1970).

17. Boettner, E., *et al. Journal of Polymer Science, 13* (1969).

18. Wogrolly, E. "Some Comments on the Behavior of Household Refuse Containing PVC in Refuse Combustion Plants," *Kunstoffe, 62,* 53 (1972).

19. Sundstrom, G. and B. Steen. "Investigation of the Emission of Hydrogen Chloride During Incineration of Household Wastes Under Admixture of PVC," report of work at Malmo, Sweden (December, 1969).

CHAPTER 14

FIELD INSTALLATIONS

DOMESTIC INSTALLATIONS

The Chicago Northwest incinerator operation is the newest and most modern of the large capacity (1,600 tons per day) incinerators in North America. It is designed to support four 400-ton-per-day furnaces, all embodying water wall construction. Fife[1] offers a very complete description of the equipment involved in the full scale operation of this new plant. Included in the design of this plant are a Martin (reverse acting, reciprocating) grate system, the first such system to be in use in the Western Hemisphere, and electrostatic precipitators for air pollution control. This plant, then, is the first precipitator-equipped, refuse-burning, water-walled furnace construction, steam-generating incinerator to be engineered in the United States. It is probable that its successful operation will be followed by the construction of several others with similar features.

The flow sheet for these furnaces shows refuse fed to the grates, then to the combustion chamber (water wall) from which steam is recovered and flue gases pass through electrostatic precipitators. A fly-ash slurry is recovered, and cleaned flue gases then pass into a stack and are vented to the atmosphere. Residue handling specifications were affected by the Martin-type stokers, which include a special residue-discharging device which accomplishes the quenching (and furnace seal) function without requiring a water-filled trough, flight-type residue conveyor. The Martin system may also be used in conjunction with pan-type or rubber-belt conveyors operating "in the dry." Drum screens are specified for installation at the discharge end of residue conveyors to salvage metals which the city may sell to private contractors.

Screen tailings, mainly composed of tin cans, are
sold to a contractor who accepts the tailings at the
plant. Screenings are handled separately by the
city.[1]

Air pollution control considerations dictated
the selection of the water-walled furnace/electrostatic
precipitator arrangement. The specified precipitator
efficiency is 96.87%, and it is also required that
paper char (considered as a segregated constituent
of the fly ash), be collected with the same efficiency.
Data on the anticipated characteristics of the ash to
be collected and the rate at which the ash would leave
the furnace are provided in terms of boiler off-gas
particulate concentration. Fife[1] comments that this
procedure fails to recognize that boilers of different
design and manufacture, with widely varying gas flow
arrangements, also demonstrate great variation in
their capability to act as dust collectors and that
probably only a stack-gas particulate concentration
requirement should be met. Location of the precipitators on the Northwest incinerator project was
established at the rear of the plant, supported from
on-grade foundations rather than on the roof. This
location was selected primarily for architectural
reasons. The refuse-burning system is required to
burn the "design" refuse (5,000 BTU/lb) at the
specified rate while producing a grate residue containing not more than 4% combustible material.
Tentative methods established by the American Public
Works Association are used in residue- and refuse-
sampling procedures.

The steam-producing facilities embody the best
known, most efficient equipment presently available
to designers and engineers. These are covered by
Fife. Because of the plant location, consideration
of water-cooled condensers was not possible, and
air-cooled condensers were specified. The boilers
are designed to produce steam at a header pressure
of 265 lb/sq in/g, and the auxiliary drive turbines
are sized for saturated steam at this pressure.
These turbines are arranged to exhaust to the low-
pressure steam system at about 10 lb/sq in/g, and
are separated into high-pressure and low-pressure
assemblies. The selection of the 265 lb/sq in/g
header pressure is based on the fact that lower
pressures would not significantly reduce the costs
of equipment and piping, whereas higher pressures
would create rapidly increasing costs. Saturated
steam is used to avoid the corrosion hazard associated
with the higher tube temperatures that would have
been required in superheaters. Because of the low

cost of steam (refuse fuel at no cost), the selection
of steam pressure did not include consideration of
turbine water rates and was, therefore, based on
equipment and piping costs only. The steam generated
is used to drive the major plant auxiliaries, such
as combustion-air and induced-draft fans, boiler-feed
pumps and refuse shredders. Studies were made to
determine whether the steam output should be used for
power generation. Under the conditions in Chicago,
the study[1] indicated that it was more economical to
use utility power than for the city to operate a small
power generating plant. The comparison was quite
close, so in localities with higher utility power
rates, the reverse could have been decided.

In the matter of boiler auxiliaries, the plant
system economics dictated the use of turbine drives
for major auxiliaries. These include combustion-air
and induced-draft fans, boiler-feed pumps and the
bulky-refuse shredder. Cranes, small pumps, air-
cooled condenser fans (on the roof) and all building
service system equipment drives are electrically
operated. Use of turbine drives for boiler auxiliaries
introduces the problem of provision for the initial
plant start and for succeeding starts after each total
shutdown of the four boilers. Initial planning by
the city indicated a desire for the flexibility necessary to make such a total shutdown each weekend if
considered desirable to do so. Accordingly, two of
the boilers are fitted with auxiliaries having dual
drives, turbine and motor. Either of these two units
may be started with refuse or utility gas as the fuel
and will produce the steam necessary to start other
auxiliaries as required. Normal operations will
probably not require the use of the motors, as the
city plans to sell the steam produced to nearby
industry where there is a continuous demand. Whatever the ultimate arrangement, the plant is sufficiently
flexible to enable operation under all foreseeable
conditions.

Another U.S. municipal incinerator installation
of interest is the Holmes Road refuse incinerator
plant operated by the city of Houston, Texas. The
plant maintains two furnaces of the continuous-feed
type, permitting continuous high burning rates under
steady state conditions. Gases leaving the furnaces
are cooled in water-spray chambers and tubular reheaters. These gases then enter flue gas scrubbers
which remove particulate matter (fly ash). Gases
pass through induced-draft fans and reheaters and out
of the stacks. Forced-draft fans supply air for
combustion under the refuse fuel bed. Over-fire fans

supply air to the furnaces through the sidewalls and at the gas discharge throat. Induced-draft fans maintain a negative pressure in the furnace for the safety of personnel. Fly ash from spray chambers and gas scrubbers is conveyed in sluice ways to a water treatment plant where the ash settles out and is removed. The installation began operating in August 1967, following a construction period of two years at a cost of $4,200,000. There are two 400-ton-per-day furnaces equipped with a three-stage travelling grate and continuous-feed stokers. Gas cleaning is by filter bed scrubbers, one per furnace, with a capacity of 160,000 cfm at 600°F, claimed to be 98%+ in efficiency. Operation is continuous--24 hours/day, 7 days/week.

Achinger and Daniels[2] give physical descriptions of incinerators studied in their work. Three of these are as follows:

(1) Built in 1966 in Western United States: 24-hour design capacity of 300 tons, two refractory-lined, multiple-chambered furnaces with inclined modified reciprocating grate sections followed by stationary grate sections. The air draft system consists of 11,000 cu ft/min forced-draft under-fire air fan and a 57,000 cu ft/min induced-draft fan per furnace with no over-fire air fan. Air pollution control system is wet scrubber, impingement on 42 12-inch-diameter wetted columns. The flyash in system goes, in slurry, to settling basins. Capital cost is $7,070 per ton/day capacity; operating cost is $4.90 per ton processed.

(2) Built in 1963 in Southern United States: 24-hour design capacity of 600 tons, two furnaces with three reciprocating grate sections followed by a rotary kiln section. The air draft system is a 25,000 cu ft/min forced-draft under-fire air fan per furnace and one 200-ft stack for natural draft. Air pollution control system is wet scrubber, water sprays and a baffle wall. Fly ash scrubber water is used for residue quenching; fly ash then settles in a lagoon. Capital cost is $8,429 per ton/day capacity; operating cost is $6.69 per ton processed.

(3) Built in 1967 in Southern United States: 24-hour design capacity of 400 tons, two furnaces with four sections of inclined reciprocating grates. Air draft system is a 20,000-cu ft/min forced-draft under-fire air fan, a 24,000-cu ft/min forced-draft over-fire air fan and a 120,000-cu ft/min induced-draft fan per furnace. Air pollution control system is multiple dry cyclones following a wet baffle wall.

Fly ash scrubber water goes to residue-quench tank,
then to a settling lagoon. Capital cost is $6,983
per ton/day capacity; operating cost is $9.36 per
ton processed.

FOREIGN INSTALLATIONS

Hishida[3] describes several one to two ton/day
industrial incinerators, including floor-level
multistage type; continuous feed rotary kiln multistage (by Mitsubishi Heavy Industries); PELL spiral-
flow, pressure type; fluidized bed type; and a
continuous disposer type (one ton/day capacity) for
synthetic high polymer waste. All of these are
either in use or under concerted development in
Japan. The incinerator for high polymer waste is
claimed to be very effective for incinerating PVC.
The PVC yields its HCl when dry-distilled at about
300°C in the absence of air. A rotary kiln is used
for this pretreatment--after the HCl is vaporized
from the resin, the balance is burned using conventional methods. The HCl gas from the rotary kiln is
passed through a multicyclone and a gas cooler, then
reacted with ammonia to yield ammonium chloride
which is collected by a dust collector. It is claimed
that no pollution problem from HCl ensues, no auxiliary
fuel is needed and corrosion is reduced. (Produced
by Takuma Boiler Manufacturing Company, a 50-100
ton/day unit would cost over 1 billion yen!)

In European practice in refuse disposal by incineration, Eberhardt[4] gives descriptions of several
types of incinerator plants. In Europe the aim is
to completely burn out the refuse, to utilize the
heat produced and to minimize air pollution as far as
possible through the use of expensive flue-gas cleaning
equipment. Some values that must be attained include
0.04 grains of dust per standard cubic foot of flue
gas, less than 3% combustible in the residue, and
less than 0.2% putrefying substances. The combination
of refuse incineration with large, power-station type
boilers is becoming the rule in installations in
Germany where there is a market for the steam at high
temperatures and pressures which incineration produces.
Installation descriptions follow.

(1) Kohlenscheidungs Gesselschaft m. b. h. (KSG)
has used a compartmented travelling grate. Feed control is achieved through a feeding belt. Increasing
burnout requirements are met by the arrangement of
three or four combustion grates and by a damming

device at the discharge end. Dust development is
limited by low air velocities in the grate slots.
The fundamental features--closed grate belt (even at
reversal points), low height of the grate slots with
large free cross section, etc.--of this grate
construction are well known.

(2) The Volund incinerator design has a feeding
and drying grate as well as a combustion grate (both
grates being designed as reciprocating grates) and a
rotary kiln for burnout. A new combustion drum
(rotary kiln) has been developed wherein it is
possible to bring, by hydraulic inclination equipment,
the air required for combustion to the waste products
inside the drum; in this manner retention time (and
thus combustion time) of the waste products can be
prolonged or shortened when required. Because of
this feature, a recombustion grate is unnecessary in
most cases. The space required is reduced by arranging
the feed grate and the drying grate in opposite directions above the drum. The connection to a waste heat
boiler remains a problem.

(3) The latest installation of the Von Roll
system was built for the Frankfort/Main refuse incineration plant. The incinerator, for a rate of
15 ton/hour of raw refuse, is combined with a corner
tube boiler for a steam output of 71,500 lb/hr at
958 psi and 932°F. The waste heat boiler is arranged
as a radiant boiler directly over the furnace. In
order to reduce the production of fly ash, the grates
run in the same direction, and the transfer heights
have been diminished considerably. For construction
reasons, no slag generator was installed. Instead,
another grate was installed, and a flue-gas return
system similar to other systems was used.

(4) The KSG radiation boiler with combined
oil-refuse firing, at Munster, is actually two plants
with throughputs of 20 tons/hour each. The boiler
plants are arranged side-by-side. The boilers are
identical and differ only in refuse firing, one having
a Martin reverse action system and the other a VKW
roller or rotating grate. Each plant has a steam
output of 275,000 lb/hr at 1100 psi and 977°F.
According to the heating value, with refuse, the
steam output is 71,500 to 77,000 lb/hr, but the full
steam output cannot be attained with refuse alone.
Some oil firing is, thus, a necessity.

(5) The Mannheim KSG-radiant boiler combined
refuse-oil firing plant is also equipped with two
units, each having an 18 ton/hour throughput.
Household and trade refuse, including industrial

refuse, at about 400 to 500 tons daily is handled, as far as possible, with a refuse-firing KSG compartmented travelling grate system. The quantity of refuse mentioned includes 50 tons of refuse from a nearby chemical plant. The lowest heating value refuse ranges between 1400 and 4000 BTU/lb.

SEVERAL INCINERATOR-BOILER
HEAT RECOVERY INSTALLATIONS
IN EASTERN UNITED STATES

Atlanta, Georgia

The Mayson plant at Atlanta, Georgia, as described by Day & Zimmerman Associates,[5] contains four International Volund rotary kiln furnaces, each followed by a two-drum bent tube boiler located in the furnace flue. The first two units were constructed in the early 1940's, followed by two additional units in the early 1950's. Flue gas temperatures entering the boilers are in the range of 1,500 to 1,800°F. The original boiler tube life was about 15 years. Steam generated in this plant is sold to a local steam heating system. The present condition of the boilers is considered satisfactory. Hand cleaning is resorted to for removing accumulated tube deposits during boiler shutdowns about every two weeks.

Miami, Florida

The No. 1 plant in Miami, Florida contains six circular hearth batch feed furnaces divided into two groups of three furnaces each. Each group of three furnaces has a long common flue with a secondary combustion chamber at each end. Each of the four secondary combustion chambers is associated with a two-pass boiler. Each boiler is provided with a gas by-pass flue with spray cooling. This plant was placed in operation in 1955. Flue gas temperatures entering the boilers average about 1,500°F. Steam generated in this plant is delivered to an adjacent hospital. The present condition of the boilers is considered satisfactory. Hand cleaning of tubes is performed about four times a year during boiler shutdowns.

Hempstead, New York

The Merrick plant in the town of Hempstead, New York contains four circular hearth batch feed furnaces divided into two groups of two furnaces each. Each pair of furnaces discharges flue gas through a common secondary combustion chamber and flue into a boiler. This plant was placed in operation in 1951, and the boilers were retubed after about eight years of operation. Flue gas temperatures entering the boilers vary from 1,200 to 2,000°F. The steam generated is used in the plant for electric power generation with excess steam going to a condenser. It is currently reported that approximately 50% of the boiler tubes are being removed to investigate the effect on tube fouling. The boilers are taken off the line on alternate weekends for inspection and manual cleaning.

The Oceanside plant in Hempstead contains two integral refractory furnace-convection boiler units with continuous feed furnaces. This plant was placed in service in 1965. The steam is used to generate electric power for use in the plant. The excess steam is used to distill seawater. Gas temperatures entering this boiler range upward from 1,725°F.

Providence, Rhode Island

The Field Point plant in Providence, Rhode Island contains two furnaces, each of which discharges into two boilers. The first two boilers and furnaces were placed in service in 1936 and retubed in 1956. The second two boilers and furnaces were placed in service in 1950. Replacement of the first two rows of boiler tubes has been maintained on a continuous basis. Combustion chamber temperatures range from 1,900 to 2,500°F. One boiler and one furnace are normally operated to provide steam for an electric generator, which in turn provides power for the local incinerator and sewage plant. Hand cleaning of the boilers is performed about every two weeks during shutdown of one boiler associated with the operating furnace.

Oyster Bay, New York

The plant at Oyster Bay, New York has two boilers, each fired by a pair of rectangular grate batch feed furnaces. These boilers operate on a controlled

by-pass system to maintain desired boiler pressure.
The steam is used for heating and electric power
generation for the plant. The plant was placed in
operation about 1956, and the boilers were retubed
in 1961 because of internal corrosion. Gas temperature entering the boilers averages about 1,200°F.
Hand cleaning of the gas side of the boiler was required about every four months. The boilers were
removed from service in 1968.

Boston, Massachusetts

The South Bay plant in Boston, Massachusetts has
three waste heat boilers. Each is installed in a
flue fed by two rectangular batch feed furnaces.
Boiler by-pass flues are provided for each furnace.
The plant was designed to supply steam to a nearby
hospital and was placed in operation in 1959. Average
gas temperature entering the boilers is approximately
1,300°F. Moderate amounts of steam are generated
for local plant use only. The steam supply to the
hospital was discontinued in 1968.

Oceanside, New York

The Oceanside refuse disposal plant is described
by Kiss and Velzy.[6] There are two furnaces of
300-ton/day capacity, each for burning refuse. They
are equipped with continuous-feed, four-section
rocking type grates, each with an area of 4'0 square
feet. The burning rate may be computed as 61 pounds
of refuse per hour per square foot of grate surface,
based on the nominal capacity of 300 tons/day. The
volume of each furnace is 6820 cubic feet, providing
a heat release of 18,300 BTU per hour per cubic foot
of furnace volume. The gases from the furnaces exhaust into a controlled circulation boiler at 1,800°F.
Each boiler has a maximum generating capacity of
85,000 pounds per hour of saturated steam at a pressure of 460 psig. The boiler has two circulating
pumps, one motor driven and one turbine driven;
either one may be used on standby. A controlled
circulation unit was selected because it is lighter
in weight and occupies less space than a natural
circulation boiler. Gases are cleaned through a
battery of multiple cyclones--24 cyclones, 30 inches
in diameter, for each furnace unit. Each furnace,
at rated capacity, and burning refuse of as-fired
heat value of 5000 BTU/lb turns out 125 million BTU

per hour--the theoretical air required for this is
625 pounds per million BTU fired. The clean, fresh
water needed for cooling, spray and steam generation
is obtained by an *in-situ* process of desalinization
of salt water, available in unlimited quantity within
400 feet of the plant. Each desalinization unit has
a steam-condensing capacity of 46,250 pounds per hour
and is capable of producing 115,000 gallons of fresh
water per day; there are four such units, one on
standby. The desalinization equipment costs $266,000.
The authors[6] state the daily power-producing facilities to be 1400 kilowatt-hours, 1175 for in-plant
use and 225 on-site but outside the plant. This
amounts to 873,000 kwh per month, worth $178,000 per
year at the normal rate for the Hempstead section of
Long Island, New York.

Norfolk, Virginia

     The Norfolk heat recovery incineration facilities
are described by Moore and Reardon.[7] Furnaces are
of water-walled construction with a three-section
Detroit reciprocating stoker, 10 feet wide by 36 feet
long overall. The effective grate area is considered
as 269 square feet, corresponding to a burning rate
of 56 pounds per hour of refuse per square foot of
effective grate area. Two furnaces of 180 tons/day
rated capacity each with a waste heat boiler capable
of producing 50,000 pounds of steam per hour--the
design basis is for refuse having a heating value of
5000 BTU/lb as fired--generate the steam. The heat
release is given as 325,000 BTU/hr/sq ft effective
grate surface. Steam produced is saturated; pressure
is 275 psig. Gas cleaning is by a battery of 24-inch-
diameter cyclone collectors for each unit. Boiler
auxiliaries are all sized for 60,000 pounds of steam
per hour per boiler. Two boiler feed pumps, each
having the capacity for one boiler, and one steam
turbine-driven pump having sufficient capacity for
both boilers, are supplied. Chemical pumps, mixing
tanks and other appurtenant equipment for treating
feed water are included; exhaust steam from the turbine is used in the feed water heater. A steam-
atomizing oil burner for heavy oil, having a capacity
to generate 50,000 pounds of steam per hour with oil
only, is installed in each unit. These can be used
to augment the steam production from refuse when and
if required. Weight and heat balance figures indicate that, for the rated design refuse load using
50% excess air, each furnace will produce steam at

the rate of 50,000 pounds per hour from refuse of the
5000 BTU heat value. This corresponds to 3.3 pounds
of steam per pound of refuse. For this steam-producing
installation, an investigation showed that steam production from refuse alone could be increased approximately 38% by the use of water-wall furnaces. Further,
the initial cost of water-wall installation with fan
and fly ash collection equipment capacities for 100%
excess air practically balanced the initial cost of
a refractory furnace and boiler installation with
auxiliaries based on 200% excess air. The total
contract price for the installation described by
Moore and Reardon was $2,135,000, of which $1,100,000
was for incinerating equipment and appurtenances.

Stephenson and Cafeiro[8] show data on 289 municipal
incinerators built since 1945. Data are presented on
205 plants for which questionnaires were completed.
Eighty-four plants, for which questionnaires were not
returned, are also listed. Replies covered 172 new
plants, 14 rebuilt plants (major construction based
on redesign) and 19 plant additions. In Appendix II
of the Stephenson and Cafeiro work, all 205 of the
incinerators are described. Data are given on each
as to location, year built, capacity in tons per hour,
furnace type and number, feed type, stoker type,
combustion chambers, chimneys, waste heat uses,
residue handling facilities and other information.
As an example of the data given, Table 14.1 shows
grate burning rates, including stoker types.

STANDARDS FOR AIR POLLUTION CONTROL
FROM INCINERATION OPERATIONS

Of the five major sources of air pollution, refuse
disposal is by far the lowest contributor. It yields
7.9 million tons of pollutants per year, or only 4.2%
of the total. Of this, the bulk of the pollutants
are due more to inadequate design and operation of
incinerators than to inherent factors in the disposal
methods themselves. As a percent of total by source,
incineration contributes 4.8% of the carbon monoxide,
0.3% of the sulfur oxides, 5.4% of the hydrocarbons,
5.6% of the particulates, and 4.1% of the nitrogen
oxides found in all air pollution. As direct quantities, air pollution from refuse disposal in 1966
amounted to 4.5 million tons of carbon dioxide, 0.1
million tons of sulfur dioxide, 1.4 million tons of
hydrocarbons, 1.2 million tons of particulates and
0.7 million tons of nitrogen oxides.

Table 14.1

Grate Burning Rates[8]

Stoker Type	Feed	Year to Year	Number Reporting	Maximum	Minimum	Median	Average
Manual	Batch	1946 – 1958	8	91	37.7	47/67	59
Circular	Batch	1945 – 1965	59	110	45.4	84	83.3
		1961 – 1965	2	70	70	–	70
Rocking	Batch	1949 – U.C.	37	71	32.4	57.5	56.8
		1961 – 1965	10	60	43	57/57.5	56.0
Rocking	Continuous	1963 – U.C.	9	67.5	50	58	58.7
Travelling	Continuous	1954 – U.C.	23	70	55.5	65	64.3
		1961 – U.C.	15	70	55.5	65	63.7
Reciprocating	Batch	1959 – 1965	11	87	35	57	57
		1961 – 1965	9	60	35	57	53.6
Reciprocating	Continuous	1963 – U.C.	8	75	55.5	60	62.9
Oscillating	Batch	1958 – 1964	2	69.5	60	–	64.8
		1961 – 1964	1	–	–	69.5	–

Pounds of Refuse per Square Foot of Burning Grate Surface per Hour

Table 14.2

EPA Air Quality Standards for Six Pollutants (promulgated April 30, 1971)

Pollutant	Level Not to Exceed		Comments
	Primary Standard	Secondary Standard	
Carbon Monoxide	10 milligrams/m³* (9 ppm)** 40 milligrams/m³ (35 ppm)	Same as primary Same as primary	Maximum 8-hour concentration not to be exceeded more than once yearly. Maximum 1-hour concentration not to be exceeded more than once yearly.
Hydrocarbons	160 micrograms/m³ (0.24 ppm)	Same as primary	Maximum 3-hour concentration (from 6 to 9 a.m.) not to be exceeded more than once yearly.
Nitrogen Oxides	100 micrograms/m³ (0.05 ppm)	Same as primary	Annual arithmetic mean.
Particulate Matter	75 micrograms/m³ 260 micrograms/m³	60 micrograms/m³ 150 micrograms/m³	Annual geometric mean. Maximum 24-hour concentration not to be exceeded more than once yearly.
Photochemical Oxidants	160 micrograms/m³ (0.08 ppm)	Same as primary	Maximum 1-hour concentration not to be exceeded more than once yearly.
Sulphur Oxides	80 micrograms/m³ (0.08 ppm) 365 micrograms/m³ (0.14 ppm)	60 micrograms/m³ (0.02 ppm) 260 micrograms/m³ (0.1 ppm)	Annual arithmetic mean. Maximum 24-hour concentration not to be exceeded more than once yearly.

*cubic meter.   **parts per million.

The Environmental Protection Agency (EPA), in data promulgated in April, 1971, lists standards it expects the environmental people in each state to adopt within a reasonable time. These data are shown in Table 14.2. More recently, *Federal Register* for August 17, 1971, Part II (paragraph 466.32) gives a standard for particulate matter as follows: "No person subject to the provisions of this subpart shall cause or allow the discharge into the atmosphere of particulate matter in the effluent which is in excess of 0.01 gr/scf (0.23 gr/NM^3) corrected to 12% carbon dioxide, maximum 2-hour average." Nowhere in the literature on emission standards could any mention of limits for hydrochloric acid gas from incinerator or other furnace stack be found.

REFERENCES

1. Fife, J. A. "Design of the Northwest Incinerator for the City of Chicago," paper prepared for 1970 National Incinerator Conference (ASME) at Cincinnati (May, 1970).
2. Achinger, W. C. and L. E. Daniels. "An Evaluation of Seven Incinerators," paper prepared for 1970 National Incinerator Conference (ASME) at Cincinnati (May, 1970).
3. Hishida, A. "Waste Disposal," special feature in *Japan Plastics Age*, pp. 21-32 (November, 1971).
4. Eberhardt, H. "European Practices in Refuse and Sewage Sludge Disposal by Incineration," paper prepared for 1966 National Incinerator Conference (ASME) at New York (May, 1966).
5. Day and Zimmerman Associates. "Special Studies for Incinerators in Washington, D.C.," Public Health Service Publication No. 1748, Office of Solid Waste Management Program, Environmental Protection Agency.
6. Kiss, E. W. and C. O. Velzy. "Oceanside Refuse Disposal Plant Illustrates Boiler Application for Improved Economy," paper presented at National District Heating Association, 5th Annual Meeting (June, 1965).
7. Moore, H. C. and F. X. Reardon. "A Salvage Fuel Boiler for Maximum Steam Production," paper prepared for 1966 National Incinerator Conference (ASME) at New York (May, 1966).
8. Stephenson, J. W. and A. S. Cafeiro. "Municipal Design Practices and Trends," paper prepared for 1966 National Incinerator Conference (ASME) at New York (May, 1966).

CHAPTER 15

INDUSTRIAL INCINERATION

GENERAL CONSIDERATIONS

As in most processes, the largest plant operating
on a continuous basis will be the most economical to
build and operate. In view of this, local industries
and/or municipalities should look into the possibilities
of joint ownership and/or operation. This, of course,
poses other problems, but the resulting savings can
be significant. For the individual operation which
does not wish to combine with others in the disposal
of its wastes, Woodruff and Wene[1] give several
principles which merit considerations.

The first thing to do when waste becomes a problem
is determine its origin and try to eliminate it, or
at least reduce the quantity.[1] After this step, an
investigation should be initiated to determine if
the material can be economically salvaged or used
as--or in--a by-product. Upon determining that a
disposal problem does exist, disposal methods must
be considered.

Incineration as the method of waste disposal may
be selected for several reasons:

- It may be the most economical process available.
  This could be especially true if heat recovery is
  possible, or if the fuel value of the waste material
  is sufficient to sustain combustion.
- Toxic materials which cannot be treated in any other
  fashion can be oxidized or altered to a less toxic
  state during combustion.
- Reduction in volume of the waste materials may be
  required because of limited available land area.

Sources, quantity and characteristics of all waste
materials should be identified. These can include:

- production facilities
- office operations
- service areas--careteria, metal-working shop, paint shop, maintenance, etc.
- laboratory facilities, including pilot plants
- salvage operations, such as oil recovery

The minimum, maximum and average waste production in all categories should be known, as well as the average.[1]

During a waste survey, an essential element is to determine the nature of the waste materials under consideration for incineration. In some cases, the required data on physical and chemical characteristics will be known, but more often the information must be obtained by laboratory tests and engineering estimates. Some of the key data required are: density, BTU value, moisture content, ash content, chemical reactivity, elemental analysis (spectroscopic or other method), vapor pressure, corrosive nature, viscosity, physical form (powder, pellet, sticky solid, stringy mass, etc.), solids concentration in liquids, approximate size of largest particles, metals present (elements and amounts), halogens (elements and amounts), and presence of other materials that would produce products of combustion contributing to air pollution or which are considered damaging to refractory materials.

In consideration of the criteria for selecting an appropriate incineration system, the following are some of the important ones.[1]

- Suitability of incineration design for the type of waste materials to be handled
- Emission quality and quantity from stacks
- Residue quality
- Labor requirements
- Maintenance requirements
- Utility requirements
- Construction costs

INDUSTRIAL INCINERATION EQUIPMENT

In industrial incineration, special care with regard to selection of incineration equipment is required for certain waste materials.[1] Such materials would include plastics, solid tars, filter cake, drummed material, etc. Often, it is desirable to burn organic residues in the same facility as the refuse. There have been several installations built

to accomplish dual fuel burning. Organic residues
can act as auxiliary fuel for "wet" refuse. However,
organic residue burning adds a complication to the
design of refuse incineration facilities.

There are basically three forms of modern
continuous-flow-through refuse incinerators. The
flow-through incinerator is the most economical and
most satisfactory from a combustion standpoint, if
the quantity of material to be burned is large enough
to justify this type of equipment. Incineration
furnaces utilizing metal grates of various types are
the most widely used for municipal refuse. Several
types of grate furnaces are in use, including travelling grates,rocker grates and drum-type grates.

The rotary hearth furnace is the second major
type of incinerator system. This type of furnace is
built with a rotating hearth, a fixed hearth and
rotating rabble arms. The rabble arms are of air-
cooled metal construction. The furnace is often
built with several hearths, one over the other.

The third type of incinerator is a rotary kiln
furnace. Burning in this type of furnace is accomplished on the refractory hearth which rotates
around an inclined axis.

For liquid refuse incineration, there are several
types of waste liquid burners in use which can be
divided into four basic types. These are:[1]

- box furnace,
- tube furnace,
- special reactor-type furnace, and
- submerged combustion or submerged exhaust burners.

The first three of these four types are usually
refractory lined but they can be constructed, at
least partially, with water-cooled metal walls.
They are suitable[1] for liquid residues with:

- reasonably high heat value,
- no components which would destroy refractory
  materials, and
- no components which represent an emission problem.

Most liquid residues can be successfully burned in
any of the four types of furnaces when proper design
and operation are accomplished. Liquid residues
which contain materials such as sodium, which causes
slagging and fluxing of ceramic refractories, require
special consideration. Some liquid residues contain
materials which will create a hazard to the environment

if emitted in the combustion gas. Special incineration equipment, usually coupled with gas-cleaning equipment, is required for such residues.

Ash handling systems, while they are more necessary on municipal incinerators, are desirable on industrial types. Ash from refuse and residue incineration is usually in the range of 1 to 10% by weight of materials incinerated but could run as high as 50% depending on the amount of noncombustibles in the charge. Modern incinerator practices result in an ash of very low organic content, suitable for high quality landfill.

Energy recovery is an important consideration in industrial as well as municipal incinerator operation. Depending on the nature of the materials to be incinerated, a positive heat balance will normally exist. Sense of economy would suggest that heat recovery should be considered. The logical approach is to produce steam. Steam has a value only if a user is reasonably nearby. In recent years, heat recovery has been practiced with increasing success. The economic benefit can represent a significant percentage of the total operating cost of the incineration facilities. Where steam recovery and sale is not practicable, due to excessive distance to a consumer, it may be economically feasible to use steam to generate electric power for use and sale. It must be recognized, however, that a waste disposal facility must be designed primarily to accomplish that purpose and not as a steam or power generating facility. Energy recovery, in general, warrants a careful examination.[1]

On utility requirements, usually two basic water systems must be provided, *i.e.*, process water system and a potable water system. Potable water supply, depending on the plant site selection, usually is not critical. Process water on the other hand, may be more difficult to obtain due to the much larger quantities. Process water is required for:

- gas-scrubbing equipment,
- sluicing ash,
- cooling,
- lubrication,
- fire fighting, and
- miscellaneous housekeeping.

The aggregate total requirements for process water can be appreciable. Facilities have been planned where water usage has been as high as 2-1/2 mgd. In

general, the quality of the process water is not
critical, and perhaps wastewater from other production activities might be utilized for this service.

On the matter of steam, there may be a requirement for space heating, heating of residues and
residue atomization. Many of the liquid residues
now wasted from industry require heat so that they
may be handled and pumped. In addition, steam is
useful for general equipment cleaning. Usage might
vary from a few hundred pounds per hour up to 10,000
pounds per hour or more for a large incineration
facility.

In regard to electricity, the connected electrical
load for an incineration facility can be appreciable,
especially if forced-draft fans and additional processing equipment such as a wood hog are required.

In the area of auxiliary fuel, provision is
generally made for use of auxiliary fuel when the
BTU balance requires it. The study of materials to
be burned will indicate the degree to which auxiliary
fuel will be required to maintain combustion, usually
in the 1500-1600°F range. In addition, auxiliary
fuel may be required to start incineration units.

On wastewater disposal, process wastewater which
originates in the various burning units will probably
contain mostly ash and dissolved solids. On an
equalized basis, pH of the wastewater might be acid
or alkaline, depending on the materials burned.
Adjustment of the pH of the waste stream may be required before discharge. One way of accomplishing
this is to utilize an equalization basin followed by
pH-controlled addition of acid or alkali. Thermal
pollution in some instances may be a problem and can
be avoided by providing a spray pond or cascade discharge. Care must be taken in a residue storage area
to insure that any residue loss does not reach a
natural water course. Storage areas should be diked
and drained to a sump which can be pumped. Surface
run-off water should be diverted away from the
residue storage area. Organic content of waste-
waters should be minimal with good combustion and
only under exceptional circumstances require biological
treatment before discharge. Sanitary wastes should be
collected separately and disposed of according to
sanitary practices. Compressed air is usually required for various uses such as purging, cleaning,
tool operation, instrumentation, and sometimes
atomization.[1]

Control Equipment for Gases
and Vapors

The topic is a complete subject in itself. A
minimum of spray chambers will be required for fly
ash control on a refuse incinerator. The best means
of air pollution control is to provide complete combustion and particulate separation in the incinerator,
This means the three "T's" must be provided for, *i.e.*,
temperature, turbulence and time. Significant
improvements have been made in recent years in design
of gas-cleaning equipment. However, efficient gas
cleaning is still costly in terms of energy expended
and first cost and will probably continue to be so
in the future.

Stack design should take into consideration local
topography and climatic conditions as well as draft
requirements to minimize the effect of infrequent
increases in emissions due to poor combustion or
failure of control equipment. Unburned hydrocarbons
and carbon monoxide from the main combustion chamber
of the incinerator, as well as the inorganic acidic
gases such as hydrogen chloride, sulphur oxides and
nitrogen oxides which arise from the incineration
process, should be minimal in any properly designed
and operated equipment. Organic vapors such as
organic acids should also be minimal, but it must be
admitted in practice that such materials are often
to be found in the flue gases.

As stated by Cory,[2] fortunately the acidic gases
may be substantially removed by the use of scrubbers.
In gas cleaning or scrubbing in industrial incineration
assemblies, the design parameters can be summarized
as follows:

- The water rate to the scrubber should be about 1 gpm
  for each 100 pounds per hour of rated daily incinerator
  capacity. This gives a water-to-gas ratio of about
  1 gpm per 400 scfm.
- The exhaust fan should be designed to handle 700 cfm
  at 350°F (177°C) for every 100 pounds per hour of
  incinerator capacity.
- The fan should be designed to provide 0.5-inch static
  pressure for a 50-pound-per-hour unit. These static
  pressures shoudl be developed with the fan operating
  at 350°F.
- The horsepower requirements of the fan should be
  based on the full capacity of the fan at ambient
  temperature.
- Attention should be given to materials of construction. Certain laminated or glass-filled plastics
  and high heat resisting materials are now available.

It is desirable to have a scrubber that is neither
longer nor higher than the incinerator dimensions.

In considering the overall design of scrubbers,
it has been determined that air dilution of the gases
from the incinerator prior to entering the scrubber
is unnecessary. Water should be introduced into the
effluent as it enters the scrubber and flows con-
current down its first pass. The best location for
a scrubber is at the rear of the final combustion
chamber. By immediately introducing water into the
gas stream, more time is allowed for mixing and
evaporation while the desired cooling is accomplished.
The average velocity of the gas-water mixture in the
first pass should range from 9 to 10 ft/sec. The
velocity of the gases in the up-pass is determined
by calculating the remaining time requirement so that
the gases remain within the scrubber for a total time
of approximately 1-1/4 sec. The curtain-wall port
is sized in the velocity range from 18 to 20 ft/sec
to prevent excessive pressure drop and to prevent
water in the sump from being re-entrained in the
effluent. The gases exit from the extreme top of the
up-pass so that its full length can be used for the
evaporation of any water remaining in the gas stream.
This location also prevents water, which tends to
travel up the back side of the scrubber, from becoming
re-entrained in the gas stream. Another provision,
which also reduces the possibility of re-entrainment
of the larger diameter water droplets, is the inclu-
sion of a 4-inch channel at the bottom of the curtain
wall on the down-pass side. This channel prevents
water that impinges on the wall from forming larger
diameter droplets and dripping from the bottom of the
curtain wall into the high velocity gas stream below.
The channel collects the larger droplets and carries
the water across the width of the scrubber, down its
side walls, and into the sump.[2]

Industrial Incinerator Design Principles

Much data on design principles can be found in
such references as Woodruff and Wene,[1] Corey,[2]
American Public Works Association,[3] Hescheles,[4]
Monroe,[5] Lucier,[6] Novak,[7] Zinn and Niessen,[8] Warner,
Parker and Baum.[9]

Multiple Chamber Designs

The early designs of single-chamber incinerators
have been outmoded, largely because of failure to

meet air pollution restrictions. The empirical design relationships evolved from incinerator investigations and tests have resulted in the development
of two basic types of multiple-chamber incinerators;
namely, the *retort* type and the *in-line* type. The
emission of solid and liquid particulate materials in
combustion effluents from multiple-chamber incinerators is primarily a function of the mechanical and
chemical processes taking place in the ignition
chambers.[1-3] Designwise, the fundamental relationships to be considered in calculating primary
combustion chamber parameters are length-to-width
ratio, arch height, and grate loading. Engineers
have evolved formulas governing ignition chamber
design postulated from data on tests of units of
varying capacities operated at maximum combustion
rates.[2] In application of such design parameters,
incinerator design factors that are considered most
important include the ratios of combustion-air
distribution, supplementary draft and temperature
criteria, ignition chamber dimensions, arch height,
grate loading, secondary combustion stage velocity
and proportion factors.[2,3]

In a comparison of the retort design and the
in-line design, the essential features of the *retort*
type are:[2]

- The arrangement of the chambers causes the combustion
  gases to flow through 90° turns in both horizontal
  and vertical directions.
- The return flow of the gases permits the use of a
  common wall between the mixing and secondary combustion
  chambers.
- Mixing chambers, flame ports and curtain-wall ports
  have length-to-width ratios ranging from 1:1 to 2.4:1.
- Bridge wall thickness under the flame port is a function of dimensional requirements in the mixing and
  combustion chambers. This results in construction
  that is somewhat unwieldy when capacity exceeds 500
  pounds per hour.
- Retort incinerators are rated as more efficient when
  the capacity range is low--that is, from 50 to 740
  pounds per hour.

The distinguishing features of the *in-line* design are:[2]

- Flow of the combustion gases is straight through the
  incinerator, with 90° turns only in the vertical
  direction.
- The in-line arrangement of the component chambers
  gives a rectangular plan to the incinerator. This

style is readily adaptable to installations that
require separated spacing of the chambers for
operating maintenance.

- All ports and chambers extend across the full width
of the incinerator and are as wide as the ignition
chamber. Length-to-width ratios of the flame-port,
mixing chamber, and curtain-wall port flow cross
sections range from 2:1 to 5:1.
- In-line designs function most efficiently at capacities in excess of 1000 pounds per hour. Here,
grates are long enough to maintain burning across
the full width of the ignition chamber. This results
in more satisfactory flame distribution in the
flame-port and mixing chamber.

The *open-pit incinerator*, developed at DuPont by
Monroe[5] is, in effect, a single-chamber industrial
incinerator. It was originally developed several
years ago as a disposal method for high flame-
temperature material such as nitrocellulose. The
open-pit type of incinerator, as now manufactured,
will burn many types of high BTU materials such as
polyethylene, polypropylene, nylon, and other plastic
wastes which range up to 20,000 BTU per pound. It
will burn plastics, rubber wastes, wood wastes and
numerous types of manufacturing process wastes. Some
materials have to be avoided, such as styrene plastics
which emit heavy, black smoke, and chlorinated hydrocarbons and other chemicals which yield corrosive
off-gases.[6]
The original standard size of the open-pit incinerator was 16 feet in length by 9 feet in width by
10 feet in depth. Combustion air requirements are in
the order of 1100 scfm per foot of pit length. Early
models were supplied without under-fire air, but it
has since been determined that the overall performance
and flexibility--in terms of the range of materials
it can handle--are improved through the addition of
under-fire air at nominal incremental cost. The
standard design, according to Lucier,[6] now has both
over- and under-fire inlets and the 17-foot design
has a 75 hp blower drive with flows, pressure and
velocities variable according to use. As mentioned
by Lucier,[6] Thermal Research and Engineering Corporation estimates that DuPont has twenty units in
operation and there are perhaps twenty elsewhere, so
the open-pit concept is considered to be beyond the
experimental stage. Since it is not designed to
collect and run off gases through gas cleaning equipment, the open-pit incinerator is best located where
any emitted fly ash and other noncombusted solids do
not violate emission codes.

INDUSTRIAL REFUSE

For industrial incineration, while many of the
principles are the same as those applicable to
municipal methods in determining refuse composition
and physical character, the types of refuse differ.
For example, industrial wastes may run the gamut
from solids through sludges and "goop" to just plain
liquids. According to Hescheles,[4] wastes produced
in an industrial plant can be classified into
several categories:

- Garbage (cafeterial wastes)
- Refuse (office and packing discards, etc.) such as
  office records; office correspondence; office com-
  puter cards, tapes, etc.; packing cartons; packing
  crates; pallets (wood); demolition debris
- Process wastes: not saleable and not reworkable;
  from process cleaning operations

The first two categories, garbage and refuse,
can be disposed of in the same manner as practiced
in regular municipal disposal plants. The last
category, industrial process wastes, usually requires
a specially designed installation for disposal of
industrial wastes without causing pollution--they
are not disposed of in municipal facilities. Types
of industrial wastes considered for ultimate disposal include the following: monomers, polymers and
resins; solvents; waste oils--mineral, cutting; oil
sludges; paint sludges; oil-water emulsions;
chlorinated hydrocarbons and solvents; industrial
solid wastes, trimmings, etc.; phenols, cresols,
tars; combustible chemicals; amines; fats, vegetable
oils and greases; miscellaneous.
Investigations of available equipment to burn
industrial wastes indicate the desirability of using
two furnaces, one to burn solid wastes and sludges,
the other to burn liquid wastes. A facility to
operate in this manner, picking up wastes from a
variety of industrial plants (chemical, petroleum,
etc.), and hauling to a regional plant for ultimate
disposal, is described in reasonable detail by
Hescheles.[4]
In many cases, depending on plant location, type
of refuse to be disposed of and other factors, management might be well advised to compare the cost of
disposal by hauling to a regional plant operated by
an outside contractor versus the capital investment
required and operating costs of disposal plants on
their own premises. The *regional facility* will very

frequently be able to assume an investment in equipment and other expenditures that simply cannot be justified by a single, average-size industrial plant. Further, such enterprises are able to take the full responsibility for the operation insofar as the solid waste disposal is concerned.

While plastics industrial solid wastes are estimated to be less than 12% of the total of all plastics solid waste generated--in 1969 estimated at less than 500,000 tons--some plastics producing companies may wish to dispose of their own wastes. Incinerators built to consume these relatively small quantities may be difficult to justify economically unless there are other types of wastes to be disposed of concurrently.

There are several types of incinerators which are effective for industrial solid wastes, but those which do not have easily plugged grates seem to be more widely used where plastics are involved. Rotary kiln types have been used successfully both in the U.S. and in Europe in the incineration of solid wastes containing relatively high proportions of plastic materials (up to 30% of the charge weight). Today, equipment and methods are available to treat industrial wastes for disposal without adding to either air or water pollution problems in the vicinity. Of course, residues of up to 25% of the charge are common, depending on the amount of combustible material. Provision for disposal of residue is necessary.

A regional facility is described by Hescheles.[4] The rotary kiln, or rotary incinerator, consists of a revolving refractory-lined cylinder, slightly inclined to the horizontal, supported by two riding rings resting on two trunnion rolls each, with the trunnion roll bearings mounted on a structural steel base. On each side of one of the riding rings, a flanged trunnion roll holds the cylinder in its proper longitudinal position. The cylinder is rotated by means of power-driven trunnion rolls with a through-shaft, which extends through two trunnion rolls on one side of the cylinder and is directly coupled to a helical gear speed reducer with an electric motor drive. The inside of the cylinder is refractory lined along its entire length, suitable for the required operating temperatures and for the various waste materials to be handled.

The rotary-kiln afterburner, 9 million BTU/h, is a high-speed rotary cup burner designed to burn liquid wastes and/or natural gas. The afterburner is a complete unit with its own forced-draft fan and combustion safeguards malfunction to shut off the operation. The afterburner preheats the kiln prior to startup, after which it maintains furnace temperature. The rotary kiln is equipped with furnace temperature control, temperature recorders, furnace-draft indicator, and a high furnace-temperature alarm. This is believed to be an acceptable arrangement for industrial solid waste incineration.[4]

Solid wastes are automatically fed into the furnace on a pre-programmed cycle, one batch at a time. Feed hoppers of 8-ft^3 capacity are placed on a conveyor to be unloaded into the furnace one at a time. The opening of the furnace door, unloading of the hopper, and the closing of the furnace door are all operated automatically.

Where there is a demand for such services, industrial wastes and refuse (liquid as well as solid wastes) can be collected and disposed of by companies equipped to handle large volumes of wastes generated by several factories located within reasonable transportation ranges.

The industrial solid waste portion of the plastics industry does not appear, with a few exceptions, to be large enough for most companies in this industry to justify the expense of such individual installations. Investigation of their availability on contract arrangement would appear to be of interest.

In the disposal of industrial wastes, including liquids, Novak[7] describes facilities to incinerate waste products from a large number of chemical processing plants located on 46 acres at Midland, Michigan. Plastics burned have included polystyrene (resin and foam), PVC and a number of others. Primary incineration units used for disposal of solid wastes consists of a 65 $\overline{M}$ BTU/hr rotary kiln incinerator which is used for the incineration of chemical refuse and liquid residues and an 81 $\overline{M}$ BTU/hr liquid residue incinerator. These units are backed up by one 32 $\overline{M}$ BTU/hr Hooker liquid residue incinerator, a 56 $\overline{M}$ BTU/hr Bigelow Liptak liquid residue incinerator, and a 65 $\overline{M}$ BTU/hr vertical liquid residue incinerator. All water wastes resulting from these operations are routed to wastewater treatment facilities. In addition to these facilities, also operated is a landfill area which covers 200 acres.

Solid refuse consists of plastic sheets, trimmings, strands, powders, styrene foam, paper, boxes, wood, and other solids of high BTU content, plus many nonburnable solids such as cinders, demolition dirt and sludges. The liquid residues are by-products, contaminated chemicals, or bad batches that cannot be economically reclaimed or reused. Many residues

are chlorinated and can contain as high as 50%
chlorine, plus several percent of ash in the form of
Fe, Ca, Mg, Na, oxides and chlorides.

The liquid waste materials brought to the incinerator are transferred to predesignated receiving
tanks which contain compatible wastes. The waste is
strained as it is pumped from the receiving tank
into a burning tank, where it is blended for optimum
burning characteristics. All liquid residues are
burned in suspension by atomization with steam or
air. Drum quantities of solid tars are destroyed
by feeding them into the rotary kiln incinerator
via a hydraulically operated drum and pak feeding
mechanism. All refuse, except full drums and paks
of material, is dumped into the refuse pit. An
overhead crane is used to mix the refuse and elevate
it to the charging hopper of the rotary kiln.

While refuse is being fed, liquid tars are fired
horizontally into the rotary kiln. As the refuse
moves down the kiln, the organic matter is destroyed,
leaving an inorganic ash. This ash is made up primarily of slag, plus other nonburnables such as drums
and metal. The ash discharges from the end of the
kiln into a conveyor trough that contains three feet
of water. After quenching, the material is conveyed
into a dumping trailer and landfilled.

Upon leaving the kiln, the products of combustion
enter the secondary combustion chamber and impinge
on refractory surfaces that cause a swirling action.
No secondary fuel or afterburners are used. Down-
stream of the secondary combustion chamber, the gases
pass through several banks of water sprays in the
spray chamber, in which the fly ash is knocked down
and sluiced onto the ash conveyor floor. The cooled
gases pass under a stack damper, and to a 200 foot
tall stack.

The rotary kiln does not have the capacity to
burn all of the liquid residues. Residues which have
a very low ash content are burned in this kiln, and
the remainder are routed to the 81 $\overline{M}$ BTU/hr liquid
residue incinerator. The liquid residue incinerator
has a combustion chamber 35 feet long and 10 feet
square in cross section. Residues are fed into the
unit through a combination of four dual-fired nozzles.
The combustion gases are quenched in a spray chamber,
which is followed by a high pressure drop Venturi
scrubber and a demister-cooler. About 3,000 gpm of
water is recycled from the primary tanks of the
wastewater treatment facilities to furnish scrubbing
water. The scrubbing water flows back to the

244 Solid Waste Disposal

wastewater plant for treatment. This is part of the
continual program to reuse water.[7]

For the incineration units, an effort is made to
completely define the system's operation under certain
specific conditions by actual field tests and measurements. With this knowledge, one can confidently select
the optimum operating conditions for disposal of a
variety of materials that have widely diverse properties. Another area of concern is materials of construction, as it is in any incineration operation.
Accurate operating and maintenance records are vital
to the evaluation of various materials of construction.
The selection and incorporation of reliable instruments
and controls also play a significant part in the total
efforts.[7]

INDUSTRIAL INCINERATION COSTS

Proper operation of incineration facilities will
require a reasonable amount of technical supervision
and administration and cannot, in general, be left
in unskilled hands. In some large manufacturing
facilities, it becomes important to allocate the
cost of the operation to the facilities where waste
materials originate. This may involve a rather
complex cost-splitting formula which takes into
consideration differences in BTU value, required
materials of construction, transportation, heat requirements during storage and additional processing.
Back charging of disposal costs to the originating
facility encourages reductions in waste and allows
economic comparison for proposed process improvements
which may reduce wasteloads.[1]

Unit Costs

Construction costs for industrial incineration
facilities tend to be somewhat higher than for
municipal facilities of comparable capacities, as
most industries try to minimize manpower requirements
and their waste materials often require a more sophisticated approach. In addition, industry tends to be
more concerned with maintenance costs than are
municipalities and therefore tends to buy systems
with lower operating costs, but often of high initial
cost. For a particular situation, construction costs
could be estimated for the required incineration
facilities. To this construction cost should be
added project costs. These include engineering costs,

purchasing costs, labor, hiring and training of
personnel to operate, start-up, etc.

The unit cost for disposal for a particular waste
material in industrial incineration should represent,
as much as is practicable, the actual costs involved
in its handling and disposal. Unit charges could be
made up of:

- Basic unit charges, which cover (a) indirect costs
  such as capital charges for engineering, financing,
  land, etc., and administrative costs, (b) direct
  costs including capital, labor, maintenance,
  chemicals and utilities associated with the disposal
  of industrial waste materials.
- Supplemental unit charges which cover special
  handling; preparation, processing before incineration;
  special materials of construction; combustion gas
  scrubbing and associated energy and water treatment
  costs; adjustment of BTU value to cover auxiliary
  fuel costs or heat recovery, if any.

Unit charges should be adjusted periodically, as
necessary to reflect variations in load and types
of industrial wastes being processed.[1] The above
discussion indicates the various areas which contribute to total costs, initial and operating.

Cost of Multiple-Chamber Incinerators[2]

Many of the factors that affect the cost of most
commodities are related to the cost of multiple-
chamber incinerators. This applies specifically to
commodities or services that are associated with the
building trades. Usually the cost of the incinerator
is not affected greatly by the type of materials used
in the construction of the exterior walls, whether
these be brick or steel, or in the inside linings of
the incinerator, or whether castable refractories or
firebrick is used. One exception to this is in the
use of plastic firebrick, which may increase installation costs by as much as 40%. Usually plastic
firebrick linings are installed only for high-
temperature service in multiple-chamber incinerators
when the gross heating value of the refuse is expected to be more than 8500 BTU/pound. Equipment
and installation costs are increased considerably
by the addition of such appurtenances as mechanically-
stoked grates, continuous ash-removal systems, and
automated feeding and charging mechanisms. Devices

such as these represent as much as 40% of the cost
of an incinerator. Table 15.1 shows a comparison
of the costs of multiple-chamber incinerators of
various capacities. The figures do not include the
cost of foundations, electrical wiring, and piping--
gas and water. Actual cost should be within 10% of
the basic cost of the same size incinerator anywhere
in the United States for units up to 1000 pounds per
hour capacity.

Table 15.1

Approximate Costs of Multiple-Chamber Incinerators[2]

Incinerator Capacity (lb/hr)	Incinerator Actual Cost ($)	Calculated ($/pound/hr)
50	1,200	24.0
100	1,700	17.0
150	2,000	13.3
250	2,700	10.8
500	6,000	12.0
750	9,500	12.7
1,000	12,500	12.5
1,500	20,000	13.3
2,000	25,000	12.5

Other Considerations in Costs of Industrial Incineration Installations

As discussed by Zinn and Niessen,[8] increasing
collections and disposal unit costs plus increases
in refuse generation rates have led many industries,
shopping centers, hotels and other similar under-
takings to attempt the disposal of their own solid
wastes. In many cases, the savings from less refuse
storage area and less frequent refuse collection,
together with the elimination of the unaesthetic and
unhealthful accumulation of waste appear to be sufficient to justify the installation and operation of
a private disposal system. Some of the performance
requirements such a system should meet include:
ability to consume a wide variety of waste materials,
such as plastics, wood, glass, metal foils, wire,
paper, cardboard and assorted refuse of high moisture
content--garbage, food wastes; adequate air pollution
control system provided to meet stringent codes

regarding particulates and odor control of flue
gases; relatively low capital and operating costs
for the total system; high combustion efficiency--
high disposal capacity per square foot of floor
area; ease of handling of ash residues; minimum
maintenance and no special techniques requiring
operator attention.

Industrial Incinerator Economics

A privately owned and operated incineration
system[8] may not, necessarily, represent a profitable
alternative to the use of an outside disposal (non-
public) service. It is considered of value, however,
to review the economics of incinerator operation in
order to have a cost framework concept. To this end,
figures showing the capital cost of on-site incinerators with and without scrubber are brought out in
Table 15.2.

Table 15.2
Incinerator Capital Cost[8]

Capacity (lb / hr)	Installed Cost, Unit with Scrubber	Dollars/pound/hour Unit without Scrubber
300	19	12
500	14	9
1000	8	5
2000	4.5	2.5

From these data, it can be seen that (a) below
abouve 300 lb/hr, almost any incinerator operation
becomes uneconomical; (b) economics favor larger
incinerators operating fewer hours--that is,
operating costs outweigh capital costs; (c) only
with difficulty can the marketing of incinerators
be based on the economic benefits of waste disposal
alone, without consideration of side benefits;
(d) large incinerators could be more costly and
still be economically justified if residue weight,
maintenance and manpower needs could be reduced and
unit life increased; (e) the impact of possible
stringent air pollution codes on incinerator adjunct
equipment economics is quite significant, particularly
in small units.

The work of Warner, Parker and Baum[9] gives additional detail from literature studies on industrial incineration.

REFERENCES

1. Woodruff, R. H. and A. W. Wene. "General Overall Approach to Industrial Incineration," paper prepared for 1966 National Incinerator Conference (ASME) at New York (May, 1966).

2. Corey, R. C. "On-Site Incineration of Commercial and Industrial Wastes with Multiple-Chambered Incinerators," In *Principles and Practices of Incineration,* R. E. George and J. E. Williamson, ed. (New York: J. Wiley and Sons) pp. 106-162.

3. American Public Works Association. *Municipal Refuse Disposal,* 3rd edition. (Interstate Printers and Publishers, 1970).

4. Hescheles, C. A. "Ultimate Disposal of Industrial Wastes," paper prepared for 1970 National Incinerator Conference (ASME) at Cincinnati (May, 1970).

5. Monroe, E. S., Jr. "New Developments in Industrial Incineration," paper prepared for 1966 National Incinerator Conference (ASME) at New York (May, 1966).

6. Lucier, T. E. "The Pit Incinerator," *Industrial Water Engineering,* (September, 1970).

7. Novak, R. G. "Comprehensive Industrial Solid Waste Program at Dow Chemical Company," paper presented at National Industrial Solid Waste Management Conference, University of Houston, Houston, Texas (March, 1970).

8. Zinn, R. E. and W. R. Niessen. "Commercial Incinerator Design," paper prepared for 1968 National Incinerator Conference (ASME) at New York (May, 1968).

9. Warner, A. J., C. H. Parker, and B. Baum. "Solid Waste Management of Plastics," research study for Manufacturing Chemists Association (December, 1970).

CHAPTER 16

THE FUTURE ROLE OF INCINERATION

GENERAL BACKGROUND

In articles prepared for *Waste Age* magazine, McGaughey,[1] Stephenson[2] and Cohan[3] offer opinion and observation concerning the future of incineration. Samplings of each author's thoughts are reprinted by permission of *Waste Age*.

McGaughey[1] looks at the role of incineration based on the concept of maximum recycling of resource values present in the residues of prior resource use. Some of his observations:

Because present conditions favor metals as more completely recoverable than organic matter, incineration would take on the characteristics of an industrial process rather than a waste disposal procedure. Should the material objectives of resource conservation take the direction of strict enforcement, it is conceivable that incineration at temperatures customarily used today would no longer be a process for waste disposal, and incineration at high temperatures might not be practiced at all except perhaps in a few instances where the destruction of plastics or other materials might seem more desirable than long distance hauling to a landfill site. However, as McGaughey mentioned, it might be doubtful that an area overpopulated to such a degree that landfills must be made hundreds of miles away will retain much ability to tolerate the environmental effects of municipal incineration. The cost of eliminating such effects might then be so great that long distance hauling becomes economical, at least for cities where the logistics of moving refuse to the resource stockpile are not insurmountable.

When to incinerate is intimately involved with the stockpiling of resources for which no present

use exists, or which must await a more appropriate
time to be economically feasible. Under this concept
only those fractions of the "solid waste" which are
needed at this time might then use incineration. A
typical example of incineration as a segregating process might perhaps be the recovery of metals from
containers and mixed refuse from which paper has been
removed for recycle. Carried to its upper limit this
might involve incineration essentially to the same
degree as it is practiced today. Only the objective
of volume reduction and materials stability would be
replaced by one of salvage with a cautious decision
to sacrifice some possible resource values in the
waste stream. Carried to the lower limit where all
materials, with the possible exception of brickbats
(broken-up concrete and soil), are considered as
resources for the future, incineration would have no
immediate role in refuse segregation and the time
when it might be applied would be postponed until
the indefinite future when the question of recovering
resources from the stockpile must be considered. In
the larger span, McGaughey suggests it is not incon-
ceivable that people might come to demand a conservation of resources with the same uncompromising
insistence that currently characterize the water and
air pollution control programs.

Stephenson[2] considered the future of incineration
from technological and other standpoints. Some of
these thoughts and opinions include the following:

The perfect incinerator plant has not been
built or designed either here or abroad, but the
nature of most of our outstanding problems is well
understood and documented. Considering the rate at
which incinerator technology has improved during the
past decade, it seems we must be on the threshold of
overcoming many, if not most, of the remaining major
problems. The introduction of European stokers to
America now gives the designer a wider selection
from which to choose the grate most appropriate for
each plant. In addition, a new 600-ton/day waterwall
steam producing plant planned for Hamilton, Ontario
will shred all incoming refuse and will utilize a
method of "burning in suspension." Little has been
published on incinerator wastewater, but progress
suggests it is amenable to established wastewater
treatment practices and, when faced realistically,
should not present a serious problem in the future.

Waste heat utilization still presents physical problems, even in the most sophisticated plants. Slugging, erosion and corrosion of boiler tubes are among the foremost. However, these problems are receiving careful scrutiny, both here and abroad, and early solutions can be anticipated.

One immediate problem lies in the skyrocketing cost of incinerator construction, a cost which may be increasing even more sharply than construction in general. Air pollution control is almost becoming "the tail that wags the dog," but certain economies are and will be possible, both in construction alone and in the overall cost of incineration construction and operation. Within a short time, sufficient data should be available on American experience with high-efficiency air pollution control equipment-- particularly electrostatic precipitators and medium energy scrubbers--to provide a sound basis for selection of equipment to be included in a plant. Investigation during the planning stage to determine whether savings in air pollution control equipment can help justify a waste heat boiler installation seems warranted, particularly for larger plants. An area with great potential for reduction in construction cost is the structure which now almost without exception completely encloses the plant.

Modern instrumentation, controls and mechanization have vastly reduced personnal requirements for incinerator operation. While the man-hour requirements have decreased, the unit cost of construction, maintenance, supplies and utilities has greatly increased for smaller plants. Recent investigations have shown that overall costs, including debt service, can be as high as $30 per ton processed for a 50-ton plant operated one shift a day. Economic reason demands that an area of many small communities approach incineration on a regional or multi-municipality basis. The day when each community could afford its own 50- or 75-ton incinerator is past.

With knowledgeable application of the present technology and anticipated further improvements, modern refuse incinerators provide a means of solving the overwhelming problem of municipal solid waste disposal for many years in the future. At least 20 to 25 years of satisfactory, nuisance-free operation can be expected of a well designed plant, provided it is properly operated and maintained.

Increasingly stringent air pollution control regulations will probably result in construction of more small industrial and special-purpose incinerators.

Waste heat reclamation will continue to offer attractive possibilities, especially for larger plants in central locations, operating 24 hours a day. Experience to date indicates waterwall construction has many potential advantages over refractory linings for incinerator-boiler plants; assuming the problems of tube wastage can be overcome, this type of construction will probably be utilized in most such installations in the future.

In the opinion of Stephenson, preliminary investigation will show large-scale waste heat reclamation to be economically justifiable in relatively few installations, and that the majority of incinerator plants built in the foreseeable future will be refractory lined. In particular, waste heat reclamation is not economical in a plant operating less than 24 hours a day except for special applications or under unusual conditions.

Looking ahead, the slagging and fluidized bed processes present interesting possibilities. Units now operating in this country are experimental, and most require several years of further development before they can confidently be applied to large scale municipal refuse disposal; however, their potential for reduced construction and operating costs, production of high quality marketable residue, and possible use of the exhaust gas as a fuel, lead to the conclusion that these processes may well point the way for the more distant future.

Last, but not least, the term "incinerator" seems to have become a dirty word in some circles and is not truly applicable in all instances. Perhaps, to complete the transition from a dirty, smelly, necessary evil to a modern industrial-type plant, the time has come to drop "incinerator" and the picture of the past which it frequently calls to mind, and adopt another term which will depict the type of plant we now build and foresee for the future.

Cohan,[3] in a move away from supplying incineration equipment to detailed custom specifications and over to standardized burning systems, lays out some comments and opinions on standardization:

The present concept of supplying incineration equipment to detailed specifications sometimes precludes the very important consideration of systems engineering. This is unfortunate because incineration is a process and, as such, is a series of dependent variables. The performance of one component is, therefore, affected by the performance of others in

the chain. This means that individually selected
components may be adequate, but when they are integrated into a system an incompatible layout could
result. This results in poor plant design and performance, and the logical solution to this problem
is the system approach. By firmly fixing the
responsibility for design and selection of the
system with one group, the probability of meeting
performance goals is increased. A properly evaluated performance specification, which includes a
system concept, offers municipalities a combination
of benefits such as integrated performance and unit
responsibility. Another solution which may not be
as apparent is standardization as opposed to the
custom designing of individual units. It is recognized that considerable resistance to standardized
burning systems can be expected initially. Naturally,
options among nonproprietary components must be
evaluated in line with the objective of optimum
technical performance. However, the number of
options should be kept within limits that will prevent dissipation of the technical and economic
benefits afforded by standardization. Refuse incineration technology has reached a state where
considerable standardization of burning systems is
feasible. Standard burning system modules are being
developed now without much risk that they will be
quickly made obsolete by a basic new advance. These
systems are proving that:

- Standardization is compatible with technological
  progress. A basic burning system can be designed to
  accommodate future improvements in equipment such as
  stokers, air pollution control systems and equipment,
  and instrumental and control systems.
- Standardization is compatible with adaptability. The
  particular needs and preferences of most communities
  can be met through options to the basic burning
  system modules--for example, needs and preferences
  with respect to total plant capacity, number of
  furnaces per plant, degree of air pollution control,
  degree of automation, and type of stack (stub or
  regular).
- Standardization is compatible with future improvement
  of individual plants. Retrofitting or replacement is
  a design consideration in both the basic modules and
  the advances that are achieved in system design.

A standardized plant--built from options to
satisfy the municipality's individual requirements--
would offer a decisive combination of advantages over

a custom-designed plant. Standardization makes it
possible to lower the initial price, guarantee burning
performance and pollution control, and to reliably
demonstrate total annual operating and maintenance
costs. Present attitudes toward the idea of standardized plants, designed and supplied by manufacturers, are largely negative. It is widely believed
that past standardization practices--in the period
from the 1920's to the 1950's--were mainly responsible
for slowing development of incineration technology.
Progress in recent years is attributed to the growing
capability of city engineers and consulting architect
engineers, and their initiative in taking responsibility
for the basic design of individual incinerator plants.
    These negative attitudes toward standardized
systems have persisted largely because few companies
undertook a concerted effort to make standardization
compatible with technological progress. Those who
tried encountered initial resistance and suspicion--
and a "thousand reasons" why it won't work. Such
resistance to standardized modular plants is gradually
being overcome. To do so, a company must be able to
show that:

  • Its standardized plants have proven advantages--both
    technical and economical--when compared with custom-
    designed plants.
  • It is capable of making substantial contributions to
    incineration technology, and will undertake a program
    to do so.
  • It will incorporate proven technological advances in
    its standardized modules and, where feasible, contract
    to incorporate improvements in modules that are
    already in service.
  • Its development program will seek not only improve-
    ments to existing modules, but also fundamental
    advances to the system concept.

    Today, standardized units are available from
incinerator manufacturers. Standardization and
shop-assembly eliminate the expense of custom engineering and minimize field erection costs as compared
to those of a "brick and mortar" installation. The
lower cost makes the incinerator system available to
a wider range of customers. The prevailing attitude,
however, is that projects like incinerator plants
require a great deal of survey work, detailed
engineering, evaluation, and coordination. Some
would be opposed initially to the idea of standardized
plants. But this opposition will be quickly overcome

once the advantages and lower capital investment are demonstrated by actual installations and operating results.

COMMENTS, DATA AND STATISTICS
ON MUNICIPAL INCINERATOR CAPACITY,
1966-1975

The data shown here appeared originally in a report prepared by APWA Research Foundation.[4]

Refuse Disposal Operations

The disposal operations in a community are handled in various ways. Excluding disposal by the producers themselves, public agencies sometimes per- form the complete task with public employees; in other cases public agencies contract with one or more private, profit-making organizations; in still other instances, all waste disposal is done through agreements between the individual producer and pri- vate enterprise or through various combinations of the above three methods. A 1964 survey of 995 communities with 5,000 or more inhabitants shows the structure of refuse collection practices seen in Table 16.1.

Table 16.1

Breakdown of Refuse Collection Practices
by Type of Collection Organization, 1964[4]

Collection Organization	Percent Share of Total Number of Communities
Municipal	44.3%
Contract	17.6
Private	13.1
Municipal and Contract	3.3
Municipal and Private	15.2
Municipal, Contract and Private	1.6
Contract and Private	4.4
Unknown	0.5

It is estimated that in the majority of small
communities with less than 5,000 population, the
disposal is handled by private companies or the producers of the wastes. A limited 1966 APWA survey
indicated that about 15% of the solid wastes in
communities of more than 10,000 population were
disposed of in open dumps, about 65% in sanitary
landfills (of which only about 10% receive daily
cover material), about 18% through incineration,
and about 2% by other methods.

Incinerator Capital Plant
in the United States

As of mid-1965, it was estimated that there were
280 to 345 non-captive municipal incinerators in the
United States. Non-captive installations are those
that are not operated for the disposal of the owner's
refuse exclusively. A detailed breakdown of the
estimates on the number of municipal incinerators is
given in Table 16.2.
Based on two surveys made in the late 1950's and
the estimated incinerator building activity since
then, it is estimated that 28% of the incinerators
were built prior to 1941, about 59% during 1940-1960,
and about 13% since 1961. Non-captive refuse collection and disposal facilities are generally owned by
local governments or private profit-making organizations or individuals. None are known to be owned by
state governments, state agencies, the federal
government, or by private, non-profit organizations
and cooperatives. For the 1966 analysis the value
of the incinerators is calculated at $2,500 to $3,000
per ton of installed capacity. The value of the
collection facilities is calculated at an average of
$10,000 per vehicle plus 12%, according to a 1966
APWA survey, for equipment, storage and maintenance
facilities.[4]

Construction Costs and
Operating Costs

Construction costs per ton of incinerator capacity
were customarily estimated to range from $3,000 to
$6,000 up to about 1965. However, a 1966 survey of
eight incinerators just completed or still under
construction indicates an average construction cost
of $4,500 per ton daily capacity. Construction cost

Table 16.2

Estimated Distribution of the Numbers of Incinerators by Community Size, 1965[4]

Community Population (1,000's)	Number of Communities in U.S. 1960	Percentage of Communities With Incinerators	Average Number Incinerators per Community	Distribution of Incinerators Number	Distribution of Incinerators Percent
1,000 or over	5	80.0	4	16	5.1
500-999.9	16	75.0	2	24	7.6
250-499.9	30	50.0	1.5	22	7.0
100-249.9	81	30.0	1	24	7.6
50- 99.9	201	25.0	1	50	15.9
25- 49.9	432	10.0	1	43	13.7
10- 24.9	1,134	7.0	1	79	25.2
5- 9.9	1,394	4.0	1	56	17.9
Totals	3,293			314	100.0

increases considerably if air pollution control
equipment, automated process controls, highly
mechanized operations, and adequate storage facilities
for the raw refuse are provided. The 1966 construction costs for an incinerator utilizing the improvements available from modern technology are estimated
to average $5,000 to $7,000 per ton daily capacity.
The cost could go as high as $8,000 to $10,000 per
ton daily capacity for plants incorporating heat
recovery systems and buildings suitable for cold
climates. Investment values range from $205 million
to $246 million (at cost) for the estimated tons of
daily incinerator capacity in operation in 1965.[4]
Costs in 1970-1971 run to considerably more than
these 1965-1966 estimates.

According to APWA surveys, it is estimated that
about 35-36% of the communities finance their refuse
collection and disposal operations through service
charges, 50-52% through general taxes and 12-15%
through a combination of taxes and service charges.
The extent to which general obligation borrowings
of local governments are used for this purpose is
not known. However, it is believed that the cost
for acquiring incinerators in many cases is covered
through general obligation bond issues. Revenue
bonds amortized by service charges have also been
issued for such purposes.

Trends of Capital Outlays
in Incineration

All capital outlays, during the 1956-1965 decade,
for the establishment of refuse disposal facilities
were made by local governments or proprietary,
profit-making organizations. It is estimated that
local governments expended about 70-75% of the total
amount and private organizations the remaining 25-30%.
Expenditures by local governments are estimated to
include $170-$222 million for incinerators. The
sources of financing for these capital outlays include appropriations from tax sources, tax exempt
municipal bonds, borrowing from banks, and private
venture capital. It is assumed that almost all
incinerators (99% of $170-$222 million) were financed
through tax exempt municipal bonds.[4]

Needs and Prospective Capital Outlays

Industrial and technological changes plus an increase in living standards are resulting in the production of ever increasing quantities of refuse, per person. This increase, coupled with the anticipated population growth, results in staggering amounts of solid wastes that must be regularly collected, transported and disposed of. Conditioned upon the 1966 situation, the capital requirements for non-captive refuse collection and disposal facilities during the 1966-1975 decade are estimated to be at least $2.42 billion in 1965 dollars. This estimate is based on a survey of the capital investment needs for waste disposal facilities recently conducted by the APWA in 47 communities and the findings of the previous analyses. The amounts of these capital investment demands are estimated to be as follows: $1.420 billion for collection equipment and storage and maintenance facilities, $340 million for sanitary landfills including land and equipment, and $660 million for incinerators.

Capacity Projections

The $660 million investment needed for incinerators includes an allowance of from 3 to 5% of this amount for land acquisition. The need for replacement of obsolete facilities is estimated to amount to 40% of the 1966 installed 82,000-tons-per-day incinerator capacity. Almost 30% of the existing capacity is estimated to have been built prior to 1941. Calculated at a construction cost of $6,000 per ton, 24 hour capacity, this capital investment would add 109,000 tons of daily capacity to the present total capacity, with 33,000 tons of daily capacity becoming obsolete. Thus the 1975 installed incinerator capacity is estimated at over 158,000 tons per 24-hour operation. In support of this estimate it might be mentioned that a manufacturer of incinerator equipment forecasts, for 1975, an incinerator capacity of 120,000 to 145,000 tons per day. However, this forecast is based on a normal expansion of the demand and does not provide for stepped up federal activities in this field.

The foregoing projections appear to be reasonable in the light of the capital requirement projections made by 20 metropolitan or regional planning commissions in urban and in some urban/rural areas. These

agencies in 1966 estimated that each of them should,
realistically, spend an average of $7.5 million during
the 1966-1975 decade, on capital investments for all
refuse collection and disposal facilities. Since
there are 216 metropolitan urban areas in this country,
their total capital investment needs are calculated
at $1.6 billion. Since such areas, however, account
for about 70 to 75% of the population, the total U.S.
investment needs on this basis can be extrapolated
to $2.1 billion to $2.3 billion.[4]

Capital Investment Needs
on an Annual Basis

If the projected needs for all municipal collection and disposal operations were to be financed over
the 1966-1975 decade in equal proportions, the annual
investment would amount to approximately $240 million
per year. If the backlog, which is estimated to be
at least 34% of the investment needs, were to be
funded during the first year of the decade, about
$820 million would be required. Spreading the re-
maining $1.6 billion evenly over the ten-year period
would add $160 million to the first year's require-
ments. Thus, it would be necessary to provide more
than 40% or $980 million of the total $2.42 billion
capital investment needs during the first year. The
remaining $1.44 billion would be required at the rate
of $160 million annually during each of the remaining
nine years. Of this amount, approximately $90 million
would be required by local government, and $70 million
by private entrepreneurs. As previously mentioned,
$660 million of the $2.42 billion is in incineration,
representing an annual investment on this basis of
$66 million.[4]

ADDITIONAL DATA ON MUNICIPAL
INCINERATION FUTURES, 1970-1985

Data used here appear in work by Niessen *et al.*[5]
and also in previous discussions in the current study.
The tabular data review and summarize that previously
presented in more detailed form.

Forecast of New Incinerator and Air Pollution
Control (APC) Capacity Distribution
in the United States, 1970-1985

The distribution used is shown in Table 16.3.
It should be recognized that the values selected are
not based upon a detailed marketing study of incinerator or air pollution control (APC) system penetration of the incinerator marketplace. Rather,
they reflect, in broad terms, the anticipated
emergence of new concepts and the increasing tendency
of communities to select systems with either low air
pollution potential, attractive economics, or ex-
ceptionally good residue quality. Also inherent in
these projections is the assumption that heat recovery
will become more important in the latter parts of the
century as a consequence of the diminishing availability (and thus higher cost) of fossil fuels. The
substantial increases in solid waste load will also
tend to justify the larger plants necessary for
economical operation with steam generation. Relative
to air pollution control systems, it is anticipated
that steady movement toward better air pollution
control will occur in future years, with some shift
away from wet systems and their attendant steam
plumes toward high-efficiency, dry electrostatic
precipitators.[5] Dry bed filters are also future
possibilities.

No attempt will be made here to defend any one
of the values assumed. It is certainly true that
any individual familiar with this field could develop
an alternative set of assumptions. It is felt,
however, that the resulting emission levels would
not be greatly changed by replacement of the values
with reasonable alternative assumptions.

Economic Data

Table 16.3 also shows assumed average incinerator
furnace capacity and consequent investment and total
operating costs (excluding APC) for each of the 14
general incinerator types considered. Operating costs
are based on incinerators of the indicated average
size operating on a two-shift per day, five-day week
basis. APC investment and total operating costs are
also shown on an assumed 200 tons/24-hour day (TPD)
plant which produces approximately 100,000 cfm of
flue gas at 700°F.[5]

Table 16.3

Forecast Distribution of New Construction and Typical Economics among Incinerator and APC Concepts[5]

A. Incineration Systems	1968 Inventory* (%)	% of New Construction for Time Span				Average[b] Capacity[c] (TPD)	Investment $/TPD (D16)	Operating Costs ($/Ton)
		'63–'70	'70–'75	'75–'80	'80–'85			
1. Continuous, Refractory, Rocking Grate	7.30	23	24	20	16	225	6,400	6.16
2. Continuous, Refractory, Reciprocating Grate	4.20	19	21	18	14	225	6,400	6.16
3. Continuous, Traveling Grate	20.87	33	26	20	16	225	6,400	6.16
4. Continuous, Refractory, Grate and Kiln	7.06	13	12	15	16	300	5,900	5.75
5. Batch, Refractory Circular	24.30	3	0	0	0	130	5,350	6.50
6. Batch, Refractory, Rectangular Cell	23.13	3	0	0	0	150	5,100	7.16
7. Batch, Refractory, Hearth	13.15	0	0	0	0	75	5,100	8.56
8. Continuous, Water Wall, Rocking Grate	0.00	0	2	3	3	350	8,750	5.48
9. Continuous, Water Wall, Reciprocating Grate	0.00	2	3	3	4	350	8,750	5.48
10. Continuous, Water Wall, Traveling Grate	0.00	2	4	5	6	350	8,750	5.48
11. Continuous, Water Wall Suspension Burning	0.00	2	5	6	9	300	9,350	8.06
12. Continuous, Slagging Type[a]	0.00	0	2	4	5	175	11,800	8.83

Item	% [*]						[b]	[c]
13. Continuous, Fluid Bed	0.00	0	0	0	1			
14. All other types	0.00	0	0	6	10			
TOTAL	100.00%	100%	100%	100%	100%	60	18,400	21.23
APC Systems [d]								
1. None (flue settling only)	17.40	0	0	0	0		110	0.05
2. Dry Expansion Chamber	21.65	0	0	0	0		140	0.05
3. Wet Bottom Expansion Chamber	2.09	0	0	0	0		170	0.14
4. Spray Chamber	12.15	3	0	0	0		175	0.30
5. Wetted Wall Chamber	16.87	0	0	0	0		175	0.30
6. Wetted, Close-Spaced Baffles	15.84	40	30	20	10		215	0.32
7. Mechanical Cyclone (dry)	9.22	15	17	20	25		340	0.77
8. Medium-Energy Wet Scrubber	4.78	30	35	38	38		780	0.99
9. Electrostatic Precipitator	0.00	12	16	20	25		900	0.98
10. Fabric Filter	0.00	0	2	2	2		900	1.10
TOTAL	100.00%	100%	100%	100%	100%			

*In % of actual total tonnage in 1968.
aSlagging type I comprises a system in which a major part is maintained at a temperature of 2800°F.
bCapacity expressed in units of tons per 24 hour day (TPD)
c1969 dollars for two shifts/day operation
d200-TPD unit, 200% excess air, 600°F inlet temperature—equivalent to 100,000 cfm

Technical-Economic Considerations

Presently the total quantity of collected municipal solid waste generated in the United States is about 200 million tons per year. Some 90% of this waste is disposed of by landfill, with only 9% being incinerated. The nub of the problem is that waste requiring disposal is increasing at the rate of 7 million tons per year, while the area available for economic disposal by landfill is rapidly dwindling. An alternative, of course, is to incinerate an increased percentage of this material. Of course, the air pollution problem would grow proportionately unless new, more effective and efficient burning processes were to be adopted. Incineration is an increasingly attractive solution, and it becomes still more economically attractive as land values and transportation costs continue to rise. The fact that incineration facilities can be located near population centers and require greatly reduced land areas in comparison to landfills enhances their attractiveness to urban planners. Thus--in spite of the air pollution problems (which are controllable to a large degree)--incineration will continue as an important method of disposal in the decades ahead.

A study showed that an incinerator capacity in 1969 of some 80 to 90,000 tons per 24-hour day (TPD) was presently existing and active in the United States. Further, from charts and graphs, projections to 100,000 tons per day (TPD), 125,000 for 1975, 160,000 for 1980, and 190,000 for 1985 are made, and an *average* increase of 7,000 TPD of new capacity per year is projected for the next thirty years.[5]

Incinerator Air Pollution

For purposes of credibility, incinerator air pollution emission data were collected from a wide variety of sources and correlated, as nearly as possible, with refuse composition and incinerator design characteristics and operating parameters. Estimates of the potential pollution load, both uncontrolled (furnace emissions) and abated by air pollution control (APC) devices (stack emissions), were provided by combining the resulting emission factors with the present and future forecast of the incinerator inventory. Consideration was given to the effects of changing refuse composition. The estimates for the reference year (1968) and the projected year (2000) are summarized, by pollutant, in Table 16.4.

Table 16.4

Estimated Incinerator Emissions[5]
(thousands of tons per year)

Pollutant	1968 Emissions Estimate		2000 Emissions Estimate	
	Furnace	Stack	Furnace	Stack
Mineral Particulate	90	56	708	118
Combustible Particulate	38	32	131	49
Carbon Monoxide	280	280	829	829
Subtotal (particulates)	182	142	1064	391
Hydrocarbons	22	22	64	64
Sulfur Dioxide	32	32	161	160
Nitrogen Oxides	26	22	147	114
Hydrogen Chloride	8	6	219	147
Volatile Metals (lead)	0.3	0.3	0.055	0.025
Polynuclear Hydrocarbons	0.01	0.005	0.03	0.009
Total	496	450	2269	1481

In general, the public's increasing concern for
air pollution abatement (Earth Day 1970, for example)
in the urban environment is reflected in the air
pollution control patterns evident in the nation today.
The interest in new incineration technology points
to the potential for reversing the trend in total
incinerator emissions, particularly mineral particulate and combustible pollutants. A sizable percentage
of incinerator air pollutants cannot as yet be controlled by existing APC devices. Certain pollutants,
such as very small particulates (less than 1 micron),
and a variety of combustible gases, such as hydrocarbons and carbon monoxide, fall into this category.
Although exotic catalytic agents and unique absorbent
systems can be postulated to control such pollutants,
cost constraints point to combustion within the incinerator as more plausible. In fact, a burnout
analysis of these pollutants indicated the residence
time in contemporary incinerators, even under the
worst of conditions, to be several orders of magnitude
higher than calculated combustion times. High combustible pollutant emission thus points to inadequate
mixing within the incinerator system where the
combustible and the oxidant are prevented from mixing
adequately. Because such mixing is evidently possible
when other low-rank fossil fuels are used in industrial practice, substantial improvement in diminishing

pollution of the air from incineration should ensue
once this knowledge is properly exploited.

Estimates of total operating costs showed that
shifting to a national average of 90% particulate
removal efficiency for all new plants would approximately double the operating costs associated with
the air pollution control systems above those of
present systems. The impact on total cost of
incineration, however, would be approximately 6%.

A technical-economic analysis[5] of various aspects
of incineration technology was undertaken to (1) assist
the existing incinerator plant owner in improving the
air pollution performance of his plant, (2) offer
background information to the municipality seeking
a contemporary incineration system offering optimum
performance, and (3) point out to various municipal,
state, and federal agencies the opportunities available in new incineration technology to realize both
satisfactory incineration and minimum air pollution--
all within pertinent economic constraints. To meet
the first objective, a detailed analysis of contemporary incineration systems was made. Although
minimum air pollution is an important and increasingly
necessary constraint, the primary purpose of an
incinerator is to dispose of solid waste at minimum
cost and with maximum safety and reliability.

In general, the overall design characteristics
of most incinerator systems can meet acceptable
performance standards including those of air pollution control. However, there was an unwillingness
or inability of the municipality to reach for the
ultimate in every aspect of incineration performance.
In addition to design suitability, however, operating
practices were found to vary from plant to plant with
some organizations willing (or fortunate enough) to
employ and support the type of operator who would
effect top performance at all times, with others
merely apathetic to the general qualification of its
personnel and the state of repair of the equipment.
Naturally, in the latter case, the performance and
efficiency of the plant was found to suffer, often
resulting in substandard activity with attendant air
pollution problems.

The third and last portion of the study consisted
of an evaluation of advanced concepts for the incineration of municipal solid waste. Advanced incineration concepts, already proven in the chemical and
metallurgical fields, show promise in incineration
technology--(a) slagging incinerators wherein the
residue is melted, and (b) systems using a fluid-bed

principle. These represent improvements over contemporary incineration systems, but development of applied technologies for refuse feeding and handling, for example, will be necessary. From a cost and performance viewpoint, no breakthrough appears imminent.[5]

The State-of-the-Art in Incinerator Technology

The state-of-the-art in incinerator technology, in response to the need for larger, more efficient solid waste disposal facilities, is developing rapidly. Conservatism, however, is inherent in the municipality, reflecting in part the limited interest and financial resources of the cities in conducting exploratory research, and perhaps the political penalties of failure. Communications are improving, however, between the municipalities, the consulting engineers often involved in incinerator design, research firms, contractors, vendors, and the academic community. Still, the realization of improved technology in new plant construction is slow. Although a number of problem areas still exist in contemporary incineration systems, it is within the state-of-the-art to design and build efficient and reliable incinerator systems producing a minimum of air pollution. Interaction of incinerator designers with individuals and firms skilled in the combustion of comparable fuel materials in industrial applications (bagasse, wood, bark, etc.) can result in the infusion of needed technology to assure adequate burnout of combustible pollutants. Substantial advancements in the incinerator art will most likely arise from the investment of federal and state efforts and research funds. As indicated above, one cannot look to the municipalities for this leadership. Private organizations, although showing increasing interest in municipal solid waste disposal, are cautious as a result of the large investments required for successful penetration and participation in this market area. Federal assistance in the demonstration of existing technology from the power, chemical and metallurgical industries can speed the technology transfer needed by the incineration community. Also, new concepts can be more rapidly brought into the practiced art as a result of federal support. Such technological contributions can make substantial improvements in the national incinerator air pollution forecasts.

The increasing quantity of solid waste will
demand the construction of new incinerator capacity.
Although a number of municipalities have avoided
commitment on new capacity in recent years, this
trend will not continue. As a result, the potential
air pollution from incineration will increase substantially over the next several years. Because of
the typical location of incinerators in the central
city, this pollutant load will be concentrated in
the urban centers. Contemporary air pollution control systems can produce acceptable effluent quality,
although not without cost. A substantial reduction
in future emissions from this source can be effected
by communities assuming responsibility for air
quality. Air pollution from existing plants can be
reduced by improved operating procedures, modifications to the physical plant, or by the installation
of the most effective air pollution control devices.
The technology exists; all that is needed is motivation and commitment.[5]
    In Table 16.5 comparisons on performance charac-
teristics for eleven types of incinerator systems are
estimated.

The Air Pollution Aspects
of Municipal Incineration

    The primary function of an incinerator plant is
to economically reduce the volume and to sterilize
solid waste. Most plants have been built and operated
solely to meet this objective. More recently, heat
recov ery aspects are becoming important to designers
and owners. Too often, incinerator air pollution
control has been only a corrective action after the
fact, sometimes applied to the plant effluent. In
recent years, however, public outrage at the emission
of air pollutants has resulted in the enactment of
forceful air pollution codes and has emphasized the
plant performance objective of minimizing environmental pollution. Consequently, communities now
operating incinerators are seeking ways to minimize
emission; communities considering the installation
of incinerator plants desire assurance that air
pollution problems will not occur; and federal,
state and municipal agencies supporting incinerator
research are seeking new concepts which would promise
a minimum of air pollution.
    Effective control of combustible pollutant emis-
sions requires modification of the combustion chambers

Table 16.5

Summary Evaluation of Incineration Concepts Based on Existing Technology[5]

| | Criteria of Performance* | | | | | | |
| | Furnace Emissions | | Reli- | Residue | | | Potential for By-Product |
	Particu- late	Combus- tibles	ability	Quality	Safety	Cost	Credits
Batch Feed							
Rectangular and Cylindrical	4	1	4	2	OK	4	–
Continuous-Feed							
Rectangular Construction	3	3	3	3	OK	3	–
Horizontal-Cylindrical Construction	3	3	3	3	OK	3	–
Ignition Grate plus Burnout Kiln	2	4	3	4	OK	2	–
Volatilizing Kiln and Burnout Grate Construction	3	3	4	4	OK	2	–
3-Stage, Rotary-Kiln Construction	2	3	4	4	OK	2	–
Waterwall Construction (Grate Burning)	4	4	5	3	OK	2	Yes
Waterwall Construction (Suspension Burning)	2	5	5	4	OK	2	Yes
Semi-Continuous Feed							
Package System	3	3	3	3	OK	5	–
Rotary Grate	2	3	3	3	OK	4	–
Continuous Feed with Residue Fusion	2	4	3	5	OK	3	–

*Range: 1 = highly unfavorable; 5 = highly favorable.

and/or improvements in operating practices, directed
at minimizing their formation or maximizing their
rate of destruction. Since these operating practice
variations also affect the other criteria of incinerator performance, consideration must be given to
the overall implications of operating and design
changes which are oriented toward the abatement of
air pollution. Incinerator APC systems offer the
most effective means to reduce emission of mineral
particulate and are the only means of reducing the
stack concentration of noncombustible, gaseous
pollutants.

Modifications to Existing Plants

In considering modifications to existing plants
(or for incinerators yet unbuilt), consideration must
be given to meeting the Criteria of Performance:
reliability, safety, environmental pollution, cost
and the quality of residue. As is true of most
complex systems, changing the design basis for one
part of the plant or an important operating parameter
to solve a given problem often affects other aspects
of plant behavior. In order to assist in visualizing
these interactions, any plant modifications to
minimize air pollutant emission should be within the
overall framework of these five plant performance
objectives.

Incineration Concepts Based on
Existing Technology

This portion is addressed to the community seeking
an incinerator which promises to meet, to the fullest
extent possible, its "personalized" criteria of
performance, which are weighted to reflect unique
local needs, and recognizes the diversity of needs
from one community to another. There are often
perspectives which are neither "right" nor "wrong,"
assuming that each community honestly and realistically
evaluates its needs and responsibilities. Therefore,
the discussion of the existing art is intended to
provide a tool for decision-making rather than to
preempt the decision-making process. Meeting the
criteria of safety and reliability requires only the
application of good engineering design and practices.

Concepts Based on Advanced Technology

Currently, a number of federal, state, and municipal agencies, and private groups are seeking improved methods of solid waste incineration. In the case of incineration concepts based on existing technology, systems which adequately meet all of the Criteria of Performance are sought by these groups. Economics are treated only to the extent which is possible at the early stage in development. Many of the pertinent safety, reliability, residue quality and environmental pollution considerations applicable to these concepts are discussed in other sections herein. For readers needing further information in these areas of technical-economic considerations, the full work of Niessen *et al.*[5] is suggested as a major reference.

In the matter of percentage of total waste which is incinerated, projections will rest upon the base taken for refuse disposal in any given year. For 1968, the projected total of collected municipal waste is taken as 190 million tons; for 1970, 205 million tons; for 1975, 225 million tons, and for 1980, 250 million tons. Incineration consumed about 20 million tons or about 10% in 1968. By 1970 incinerator capacity took care of about 30 million tons or 14% of the total municipal waste. This is projected for 1975 at about 45 million tons or 20%. For 1980 the figures are projected as 58 million tons and 23%. These figures do not include any industrial refuse. The data are shown in Table 16.6.

Table 16.6

Incineration of Municipal Solid Wastes

	In MM Tons				In Percent of Solid Wastes Collected			
	1968	*1970*	*1975*	*1980*	*1968*	*1970*	*1975*	*1980*
Collected Municipal Wastes	190	205	225	250	100	100	100	100
Solid Waste Incinerated	20	30	45	58	10	14	20	23

SUPPLEMENTAL DATA ON INCINERATOR
EMISSIONS IN THE FUTURE

Data Review

Codes in Los Angeles County permit 0.2 grain of
particulate matter (0.374 lb/1000 lb of flue gas)
per cubic foot of gas at standard conditions of 60°F
and 14.7 psia. This is covered by Rule 52 and any
subsequent modification thereof. Similar limits for
particulate matter are found in Connecticut (0.4 lb/
1000 lb of flue gas); in the Bay Area of San Francisco,
California (Reg. 2, Chapter 1); and in the U.S. Department of HEW regulations for federal installations.
The state of New Jersey in Chapter XI of its Air
Pollution Control Code has proposed an even lower
standard of 0.1 grain per cubic foot of dry gas at
standard conditions for large municipal incinerators.

Rogus,[6] following studies of codes in many U.S.
localities and in western European cities, states
that gaseous pollutants with the exception of sulfur
oxides have no standards to be met nor does it seem
likely that any will be written in the next few years.
Exceptions which have been noted for gaseous pollutants include Los Angeles County (Rule 53a), where
sulfur compounds calculated as sulfur dioxide shall
not exceed 0.2% by volume, and Connecticut where,
effective October 1, 1970, the limit of sulfur compounds calculated as sulfur dioxide must not exceed
1.1 pounds per million BTU (0.985 lb per 1000 lb of
flue gas). Where European cities have standards to
meet, they are of this order.

Emission Data

In comparison with the potential air pollution,
emissions from the power, transportation, and process
industries in the United States, refuse incineration
appears as a relatively unimportant source type.
However, this can be deceiving, particularly where
incinerators are located in the inner portions of
urban areas. Here, these localized areas can be the
victims of pollution conditions if adequate control
devices are not provided. The following data for
1968--from *American City* magazine for March, 1971--
show some changes in emission amounts ascribed to
solid waste disposal. This reflects probably both
an increase in total solid waste and an increase
in the amount incinerated.

Nationwide emission estimates for the year 1968 by the National Air Pollution Administration of the EPA are for the most part higher than previously considered. For example, emission estimates have been included for forest fires, burning of coal refuse banks, and an increased number of industrial process sources. The numbers presented here should be representative of current emissions.

Table 16.7

Estimated Nationwide Air Pollution Emissions, 1968[7]
(million tons/year)

Source	CO	Particu-lates	SO_x[a]	HC	NO_x[b]	Total
Transportation	63.8	1.2	0.8	16.6	8.1	90.5
Fuel combustion in stationary sources	1.9	8.9	24.4	0.7	10.0	45.9
Industrial processes	9.7	7.5	7.3	4.6	0.2	29.3
Solid waste disposal	7.8	1.1	0.1	1.6	0.6	11.2
Miscellaneous	16.9	9.6	0.6	8.5	1.7	37.3
Total	100.0	28.3	33.2	32.0	20.6	214.2

[a] SO_x expressed as SO_2.

[b] NO_x expressed as NO_2.

REFERENCES

1. McGaughey, P. H. "The Role of Incineration in Recycling," *Waste Age*, pp. 24, 25 (May, 1970).
2. Stephenson, J. W. "Incineration Today and Tomorrow," *Waste Age*, (May, 1970).
3. Cohan, L. J. "Standardization--A Key to Successful Incineration Equipment Design," *Waste Age* (January/February, 1971).
4. American Public Works Association (APWA). "Solid Wastes--The Job Ahead," from a report by the APWA Research Foundation for the Subcommittee on Economic Progress of the Joint Economic Committee of the U.S. Congress, *APWA Reporter* (August, 1966).
5. Niessen, W. R., *et al.* "Systems Study of Air Pollution from Municipal Incineration," Vol. I (March, 1970) under Contract CPA-22-69-23 (Clearing House No. PB 192 378).

6. Rogus, C. A. "An Appraisal of Refuse Incineration in Western Europe," paper prepared for 1966 National Incinerator Conference (ASME) at New York (May, 1966).
7. *American City*, p. 12 (March, 1971).

PART III

SANITARY LANDFILL

CHAPTER 17

LANDFILL PRACTICES

GENERAL PRACTICES

As described by Ludwig and Black,[1] a sanitary
landfill is an engineered burial of solid wastes.
The problems common to open dumping do not develop.
The technology of sanitary landfills has been improved
through research to determine such factors as the
depth of compacted earth cover required to prevent
insect, rodent and other vector infestations; the
dangers of leachate reaching and contaminating ground
water; the amounts and types of gas production and
movement within the soil; the rate of settlement,
and many other factors. A sanitary landfill can be
a very desirable asset in a community since, when
completed, the site can be used for such purposes as
recreation and parking or light construction. In a
few cases where land values have increased phenomenally,
it has been possible to construct the necessary foundations to make building on fills a practicality.
Problems such as underground fires and dust nuisances
created during construction result from careless
operation and are easily minimized by careful
management.

George[2] mentions the study prepared for the
National Commission on Technology, Automation, and
Economic Progress in which it is recommended that
further research on sanitary landfill be conducted
to determine:

- Better methods for compacting refuse before placing
  it in landfill sites.
- Construction methods and building designs that will
  permit the use of completed sanitary landfills for
  residential and industrial construction.
- Improved methods for incorporating demolition and
  construction debris in landfills.

- Decomposition of compacted refuse in submarine canyons.
- Means of operating sanitary landfills in areas having high ground water levels.

EUROPEAN AND U.S. LANDFILL PRACTICES

Warner *et al.*,[3] in brief coverage of European landfill operations, state that some confusion exists in the terminology of solid waste disposal when such terms as "open dump," "landfill," and "sanitary landfill" are employed. Whereas in the United States these terms are reasonably defined and understood, such terms are not generally employed in Europe.

In the United Kingdom, for example, the term "controlled tipping" is analogous to our use of "sanitary landfill" and has been practiced for some 50 years. It is the least costly and the most widely practiced method for the disposal of domestic garbage, consisting of spreading the refuse with the obligatory layer of soil, in a suitable location where exposure to the elements causes rusting, oxidation, decay, and microbial breakdown of certain of the organic constituents such that compaction eventually takes place with a loss of individual identity.

In Sweden, a certain amount of open dumping takes place, but the largest amount of solid waste disposal is by sanitary landfill (about 80% of the total). The Swedes are currently making a systematic investigation of all of their counties to determine the best methods for collection, transportation and processing of solid waste from economic and aesthetic points of view. The general trend is towards more incineration and composting and away from landfilling.

Landfill is presently the most common method of municipal solid waste disposal. In the U.S. in 1966 it was estimated that 79% of all U.S. cities with populations of over 25,000 utilized landfill with almost 81% of the solid waste in these communities disposed of in this manner. Only 9% of solid waste was incinerated.

Very few of these sites, however, are *sanitary* landfill sites, and HEW's Hickman in an address to the Disposables Association stated that as late as 1967 only 6% of the 12,000 U.S. solid waste land disposal sites were sanitary. The American Society of Civil Engineers defines sanitary landfill as: "A method of disposing of refuse on land without creating nuisances or hazards to public health or safety, by

utilizing the principles of engineering to confine
the refuse to the smallest practical area, to reduce
it to the smallest practical volume, and to cover it
with a layer of earth at the conclusion of each
day's operation, or at such more frequent intervals
as may be necessary." EPA accepts this definition.
The result can be a golf course or playing field, or
as is becoming popular recently in the Northern
parts of the country, an elevated winter recreation
site for tobogganing or skiing.

R. D. Vaughan[4] of EPA's Bureau of Solid Waste
Management regards the sanitary landfill as the most
important disposal method. He has stated, however,
that greater effort should be given to study the
potential for ground water pollution and methane gas
production as well as to improving landfill methods.

TYPES OF OPERATIONS IN SANITARY LANDFILL

In discussing this subject, Sorg and Hickman[5]
divide sites into two major classifications: *area
landfills*, which comprise sites on primarily flat
lands such as marshland, tideland or marginal low-
land; and *depression landfills*, which comprise sites
that utilize natural or man-made depressions or
irregularities in the terrain such as quarries, sand
and gravel pits. The area landfill type is further
subdivided into categories according to the method
of operation. The "ramp variation" is probably the
most commonly used method of area landfill. The
"trench method" or "cut and cover" method is another
variation quite widely used. Figure 17.1 describes
the area method, Figure 17.2 shows the trench method
and Figure 17.3 sketches the ramp variation.

In a ramp type project, it is easier to see the
work and control the spreading of the refuse if it
is discharged at the base of the working face and
spread from the bottom up. This also has the tendency to screen the operation from the public view
and minimize the nuisance of blowing paper.

When the material excavated from a trench is
stockpiled adjacent to the site and later used for
cover over the compacted refuse, the operation is
called a "cut and cover" type of area fill. In some
cases, the excavated material may exceed the requirements for the cover needed and the extra material is
sold. The rate of excavation bears no relation to
the rate of refuse disposal, and in many cases long
parallel trenches are opened considerably in advance
of the need for refuse disposal. This is sometimes

Figure 17.1. The area method. The bulldozer is spreading and
   compacting a load of solid wastes. The scraper (foreground)
   is used to haul the cover material at the end of the day's
   operations. Note the portable fence that catches any
   blowing debris; these are used with any landfill method,
   whenever necessary.

Figure 17.2. The trench method. The waste collection truck
   deposits its load into the trench where the bulldozer will
   spread and compact it. At the end of the day the dragline
   will excavate soil from the future trench, and this soil
   will be used as the daily cover material. Trenches can
   also be excavated with a front-end loader, bulldozer, or
   scraper.

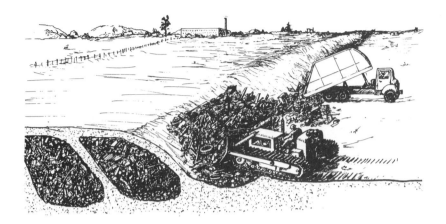

*Figure 17.3. The ramp variation. The solid wastes are being
spread and compacted on a slope. The daily cell may be
covered with earth scraped from the base of the ramp.
This variation is used with either the area or trench
method.*

a big advantage where there is excessive rainfall
or the ground may become frozen in the winter.

The cut and cover method for operating an area
fill is well suited to sites where excavation may
be made below the water table. A dragline is essential in an operation of this type and the refuse is
discharged at the top of the working face. The
dragline may be used to spread and compact the refuse.

In depression fills, the total depth of refuse
generally exceeds the depth for a single layer of
life operation. Each stratum or layer is constructed
by the placing and compacting of the refuse so that
cells are constructed with fill material on all sides
to prevent travel of fire through the mass and for
the control of rodents, flies and odors.[5]

The American Public Works Association Research
Foundation has prepared a method for determining
landfill area required.

The formula used is as follows:[6,7]

$$V = \frac{FR}{D} \left(1 - \frac{P}{100}\right)$$

where

V = landfill volume in cubic yards required for refuse
disposal per capita per year;

F = a factor which incorporates the cover material, averaging 17% for deep fills and 33% for shallow fills with corresponding F values of 1.17 and 1.33;

R = amount of refuse contributed in pounds per capita per year;

D = average density of refuse in pounds per cubic yard delivered at the landfill (about 325 for collection by compactor trucks); and

P = percent reduction of refuse volume in the landfill, varying from zero to 70%.

ADVANTAGES AND DISADVANTAGES OF SANITARY LANDFILL[5,8]

The advantages of sanitary landfill over other solid waste disposal methods are:

1. The initial capital investment is low compared to that of other disposal methods.
2. A sanitary landfill is flexible—increased or peak refuse quantities of solid wastes can be disposed of with little additional personnel and equipment.
3. A sanitary landfill can receive all types of solid wastes, eliminating the necessity of separate collections.
4. Operations can easily be terminated without a great loss in equipment or land. The equipment is of a type which can be readily used for other municipal functions and the land at any stage is no worse, and is usually better than it was before the operation began.
5. Sanitary landfill requires less land than open dumping because the refuse is compacted to between 40 and 50% of its original volume and can be deposited to a greater depth by digging ditches. Approximately one acre per year is required for 10,000 persons (seven acre feet).
6. Sanitary landfill can be established immediately upon the purchase or rental of standard digging and compacting equipment and authorization to use the land. No plant has to be built before operations can begin as is true of other solid waste reduction methods.
7. Where land is available, the sanitary landfill is usually the most economical method of solid waste disposal.
8. A sanitary landfill is a complete or final disposal method as compared to incineration and composting where residue, quenching water, unusable materials, etc. remain and require further disposal.

9. Unusual materials and bulky articles do not usually cause difficulties of operations.

10. Submarginal land may be reclaimed for use as parking lots, play grounds, golf courses, airports, etc.

The disadvantages of sanitary landfill are:

1. Large amounts of land are required.
2. Sites located outside of a city are usually under some other governmental jurisdiction.
3. Winter operations present certain difficulties.
4. Prevention of ground water pollution may be costly.
5. If the distance to the sanitary landfill is very great, the cost of transfer operations may be high.

The compacting of refuse can occur at any or all three of the stages involved in refuse disposal: (a) on site use, that is, at apartment buildings, schools, supermarkets, institutions, etc. where large amounts of refuse are generated; (b) by collection vehicles with compactor units; and (c) compacting at the disposal site where heavy equipment is run over the refuse or, as in recent cases, a special vehicle is used to provide compacting.

Another method has been developed for compacting and is currently being used in North Tonawanda, New York. A large automated vehicle with a built-in compactor moves along the landfill site digging a trench about 4 feet wide by 8 feet deep. It accepts truckloads of refuse, compacts the refuse and extrudes it into the trench. While still under pressure, earth is filled in over the refuse as the vehicle moves along so that the refuse has no opportunity to spring back. The manufacturers of this equipment claim that one machine of this nature could service a town of 80,000 to 100,000 people. This machine will be evaluated under a demonstration grant to Niagara County, New York. A similar project was initiated in 1969 in Seattle, Washington, where initial results are very promising. The economics of refuse disposal is sufficient reason for compacting. However, with stricter air pollution laws which are in some cities calling for the shut-down of apartment building incinerators, on-site compacting is becoming a necessity. Compacting is a space and time saver. Space, whether on-site, in collection trucks or at landfill sites, is at a premium. In addition, the compactor-equipped collection trucks enable collection crews to collect three to four times as much refuse than without compactors, allowing a substantial saving in labor costs.[6]

LANDFILL PLANNING

A compilation by Brunner, Keller, Reid and Wheeler,[6] presents guidelines intended to describe the best in sanitary landfill technology and indicates where research is needed to further improve this technology. The compilation has been reviewed with the assistance of a panel representing federal, state and local governments, industry, and consulting engineers.

Planning for a Sanitary Landfill

A sanitary landfill requires careful planning and design plus good supervision and operation to insure sanitary and economical disposal of solid wastes. Of these requirements, operation is probably the most important as the method in which the landfill is operated will largely determine its acceptability as a disposal method. Completed use of a sanitary landfill can best be fulfilled by planning before operation actually begins. In no way should the completed use interfere with the primary purpose of proper disposal of solid waste. Upon completion, a sanitary landfill will likely require periodic maintenance and inspection. Proper design and operation will reduce the amount of maintenance required.

Plans for solid waste disposal should be coordinated with other plans and functions of the community. Reclamation of land and development of greenbelt areas are often included in community plans. Suitable sanitary landfill sites may be reserved in a land use plan.

Environmental pollution is no respector of political boundaries. Many communities have developed intermunicipal or regional compacts to more effectively deal with the solid waste problem. Costs of equipment, land and manpower may be reduced by use of such compacts. Land, otherwise unavailable to a large urban community, may be provided by the surrounding rural communities in return for sanitary disposal of their solid waste.

A knowledge of the sources, types and quantities of solid waste to be disposed of over a 50-year period forms the basis for the selection, design and operation of a disposal system. Information on wastes handled by other disposal systems in the area is helpful in evaluating the overall disposal practices of the community. This information can be used to predict waste types and quantities that might be

expected at the proposed disposal facility should
these other disposal systems break down or terminate.
Effective enforcement of bans on open burning, on-
site burning, and open dumping will affect waste
types and quantities received at the proposed disposal
facility.

All segments of a society produce solid wastes.
The types of waste from each source are varied, but
generally the waste discarded from similar sources is
likewise similar. Residential solid waste is defined
as those wastes generated and discharged from single
family and multifamily dwellings and include both
wastes generated from within the dwelling and from
the yard and activities outside the dwelling. It is
composed of: food waste, garden waste, paper products,
metals, glass and ceramics, plastics, rubber and
leather, textiles, wood, and rocks, dirt, ashes, etc.
Residential waste traditionally represents the largest
amount of waste collected in a community. In urban
areas, public or private collectors usually transport
this waste from the dwelling to the disposal site,
while in rural areas the citizens often transport
their own wastes.

Residential solid waste varies substantially in
composition and quantity between communities. Its
composition also changes from year to year as a result of new packaging techniques and consumption
patterns. The composition and quantity of residential
waste exhibits a seasonal variation. Yard trimmings
add to the waste land in the summer, while more ashes
would be expected in the winter. Analysis of a composite of residential solid waste will show a range
of percentage for material types.

Commercial solid wastes are defined as wastes
generated and discharged from a wide variety of
activities that can be categorized as retail, whole-
sale and service establishments. Commercial enterprises can be characterized by the fact that they
sell goods and services. Some examples include
department, food, home furnishings, drug, hardware,
appliance repair stores, amusement and recreational
facilities, hotels and lodging places, restaurants,
and office buildings. It is generally acknowledged
that most commercial waste has a much higher percentage of paper products than does residential
waste; however, there are exceptions such as the
restaurant and produce market wastes which produce
a high organic waste.

Industrial waste is defined as any discarded
materials resulting from an industrial establishment.

It includes processing, general plant, packaging and shipping, office and cafeteria wastes. There is little similarity in types of wastes discarded by different industries; however, it can generally be stated that wastes from a particular industry will be similar throughout that industry. This does not imply that the composition of a particular industrial waste will remain constant, as waste composition varies with processing innovations and changes in the salvage market.

In general there are two types of industrial solid waste: (1) wastes that are peculiar to the individual industry by reason of the nature of the raw material processed and the products produced; (2) wastes that are common to all types of industry. An example of wastes peculiar to individual industries are the wastes of the ordinance industry. Those solid wastes common to all industries include shipping containers, rubber tires, broken glass, waste paper, and floor sweepings.[6]

SITE SELECTION

The site for a sanitary landfill is selected on the basis of environmental conditions, socio-political restraints, economy and other considerations. Environmental conditions should include an awareness of possible contaminants derived from decomposing solid waste.

Contaminants and Purification

Solid waste deposited in landfills degrades chemically and biologically to produce solid, liquid, and gaseous products. Ferrous and other metals are oxidized; organic and inorganic wastes are changed by chemolithotrophic and chemoorganotrophic microorganisms through aerobic and anaerobic attack. Liquid waste products of microbial degradation, such as organic acids, contribute to additional chemical activity within the fill. Since solid waste deposited in a landfill is usually a very heterogeneous mass of variable composition, accurate prediction of possible contaminant quantities and production rates is not possible at this time.

Biological activity within a landfill generally follows a particular pattern. Certain elements present in solid waste first decompose aerobically. As the oxygen supply is exhausted, facultative and anaerobic microorganisms take over. Characteristic waste products of aerobic decomposition are carbon dioxide, water, nitrate, and nitrite. Typical waste products of anaerobic decomposition are methane, carbon dioxide, water, organic acids, nitrogen, ammonia, ferrous and manganous salts, and hydrogen sulfide.

Leachate-water emanating from solid waste carries dissolved and finely suspended solid waste and microbial waste products with it. Composition of leachate is important as it may affect the quality of nearby surface and ground water. Contaminants carried in leachate are dependent on the physical, chemical and biological activities occurring within the fill, and ultimately, the properties of the solid waste.

Solid Waste Density

The number of tons to be disposed at a proposed landfill may be known from data collected by weighing solid waste delivered to another sanitary landfill or an incinerator. The daily volume of compacted solid waste can then be easily determined for either a large or small landfill. The volume of soil required to cover each day's waste is estimated from the appropriate solid waste-to-cover ratio.

Some confusion exists concerning what is meant by reported densities at landfills. Solid waste density (or field density) is the weight of a unit volume of in-place solid waste. Landfill density is the weight of a unit volume of solid waste divided by the volume of solid waste and its cover material. Apparent density is the weight of solid waste divided by the volume of solid waste and its cover material. All three methods of reporting density are usually expressed as pounds per cubic yard, on a wet weight basis. Landfill densities are reported to be in the range of 700 to 900 pounds per cubic yard, although with good compaction, 1000 pounds per cubic yard are often reached.

Larger landfill sites are preferred because various site improvements are permanent, such as drainage, leachate and gas controls, and better access roads. Heavy earthmoving equipment is needed periodically to maintain grades and integrity of the final cover material at a sanitary landfill.

Regardless of the number of sites used, firm plans and designs for disposal of solid waste over a minimum of 10 years is recommended.

Socio-Political Restraints

The society that creates the vast amounts of solid waste also influences selection of a sanitary landfill site by controlling the availability of funds and land. Mismanagement of solid waste disposal in the past has alienated the public, even though they may not know what acceptable disposal is. They do recognize the unacceptable nature of wild dumping and will tend to reject any attempt at location of a disposal site near their own properties. Because waste disposal is distasteful to most people, the importance as well as the difficulty of establishing a satisfactory disposal site is often not realized. The success of any solid waste collection or disposal system depends on the support and cooperation it receives from the public. Frequently, the public is vigorously opposed to location of a proposed sanitary landfill site; while conscious of the need for an acceptable disposal system, they may not be sufficiently informed or motivated to seek it.

Numerous objections are raised by opponents of a sanitary landfill site location, a few of which are justified. A landfill requires the operation of heavy, large, earthmoving equipment which can be noisy and disturbing to people living or working nearby. Also, truck traffic volume in the area around the proposed site is increased. But many of the alleged objections are raised by a misinformed public. Charges of odors, smoke, rat infestation, and water pollution, common to open dumping, are erroneously ascribed to any proposed sanitary landfill.

The support of the public must be obtained to insure success of the proposed disposal system. The local government must develop an information dissemination program to win support for acceptable management of solid waste and the means needed to achieve it. The program should begin early in the long-range planning stages and continue after operations begin. The media available to the solid waste manager are not limited to radio, television, billboards, or newspapers, but collection vehicles, collectors, disposal facilities, and billing receipts can also be used. Enlisting support of community organizations can do much to increase public support.

Extensive "stumping" by elected and appointed officials for support of the proposed solid waste disposal system is invaluable if the speakers are knowledgeable and have sufficient aids to help them, such as slides, films, and pamphlets.

The single most important factor for winning public support of a solid waste disposal system is an elected or appointed official who firmly believes the proposed system is acceptable and needed--a person willing to accept the challenge of developing long-range and short-term plans and to see that they are properly implemented is invaluable.

REFERENCES

1. Ludwig, H. F. and R.J. Black. "Report on Solid Waste Pattern," *J. Am. Soc. of Civil Engineers*, Sanitary Engineering Division, SA2, pp. 355-370 (April, 1965).

2. George, P. C. "Solid Waste, America's Neglected Pollutant," *Nation's Cities*. Part Two--"Land Disposal." (June-September, 1970).

3. Warner, A. J., C. H. Parker, and B. Baum. "Solid Waste Management of Plastics," study prepared for Manufacturing Chemists Association (December, 1970).

4. Vaughan, R. D. "Solid Waste Management: The Federal Role," paper prepared for Environmental Equilibrium Symposium, Houston, Texas (April, 1969).

5. Sorg, T. J. and H. L. Hickman, Jr. "Sanitary Landfill Facts," Public Health Service Publication No. 1792, Second Edition, Office of Solid Wastes Management Program, Environmental Protection Agency (1970).

6. Brunner, D. R., D. J. Keller, C. W. Reid, and J. Wheeler. "Sanitary Landfill Guidelines, 1970," Review Draft of Public Health Service publication, Office of Solid Wastes Management Program, Environmental Protection Agency.

7. American Public Works Association. *Municipal Refuse Disposal*, 2nd Edition. Interstate Printers and Publishers (1966).

8. Copp, W. R. "Municipal Inventory," Vol. I, Part 2; and Seibel, J. E. "Landfill Operations," Vol. IV, Part 1, Public Health Service Publication No. 1886, Office of Solid Wastes Management Program, Environmental Protection Agency (1969).

CHAPTER 18

DESIGN, CONSTRUCTION, EQUIPMENT AND OPERATIONS

LANDFILL DESIGN

Design of a sanitary landfill lies in the development of a detailed description of the best methodology to provide safe, economic burial of the projected amount and types of solid waste under the conditions imposed by the topography, soil and bedrock conditions, climate, and hydrology of the particular site selected. It involves the planning or outlining of all site modifications and structures necessary for handling and disposing of the solid waste in a manner that will not endanger public health or the air and water resources of the area. The sanitary landfill design should be sufficiently comprehensive to provide for specific use of the site after landfilling is completed, and the capital costs and projected operating costs over the estimated life of the project must be spelled out.

The final plan or design for a sanitary landfill should describe: various facilities provided; how the site will be operated; the potential for pollution and methods for its control; the planned use of the completed sanitary landfill; and cost estimates for using the proposed site for solid waste disposal.[1] A map, to scale, showing the location of the sanitary landfill site with respect to the area to be served is necessary. A topographic map showing the landfill site and 1000 feet of the surrounding area should be provided at different stages of filling (start-up, intermediate lifts, and completed). The topographic map or other plan drawn to a scale and contour sufficient to detail the operation, should show the location of:

1. roads (on-site and off-site)
2. fencing

291

3. drainage (natural and constructed)
4. structures
5. scales
6. utilities
7. landfilled areas
8. sequence of filling
9. borrow areas
10. fire protection facilities
11. gas control devices
12. leachate collection and treatment facilities
13. rainy weather disposal area
14. entrance facility
15. landscaping
16. completed use
17. nearby water sources
18. nearby structures.

Detailed information, including the basis for design, should also be provided for:

1. gas and leachate collection and treatment (or lack of)
2. roads
3. structures (emphasis on foundation and gas venting)
4. drainage (or lack of)
5. use of completed site
6. cell construction
7. cover material needs
8. landfill equipment
9. filling method
10. daily operation.

An estimate of the capital and operation costs over the life of the landfill should be made. Information on sources of funds, equipment costs, manpower requirements and costs, land costs, and financing charges should be included in the final design report.[1]

LANDFILL CONSTRUCTION AND OPERATION PLAN

The best possible design of an engineered structure or process is of little value unless it is constructed and operated accordingly. This is especially true of a sanitary landfill as it is continually being constructed until no more solid waste is treated at the site. The daily operation of constructing the sanitary landfill according to the design should be clearly described in the operations plan.[1]
An operation plan also facilitates continuity of the sanitary landfill activities, particularly as

landfill personnel change. New supervisors and
elected officials responsible for solid waste disposal
must know what is being done at the landfill and why.
New personnel often desire to instigate changes in
the method of operation. With an operations plan in
existence, they will be able to review and evaluate
current practices before recommending alternations.

To be effective the operations plan must be sufficiently flexible to permit revision when necessary.
Revision is aided by interchange between operating
and supervisory personnel on problems concerning
operational methods. By reviewing these reports and
discussing them with the reporter, the supervisor
will be in a better position to recommend or dis-
approve proposed changes in operations.

The hours a disposal site operates depends largely
on when the waste is delivered to the site. Except
in large cities, where they may operate 24 hours a
day, most collection systems operate on an 8-hour
day, 5 or 6 days a week, delivering waste to the
disposal site throughout this time period. The
sanitary landfill should be closed except when
operators are on duty.

Weighing of the solid waste delivered to a sanitary landfill is necessary for operational and
quality control of the fill. By knowing the weight
of solid waste, the weights and volumes of cover
material used and the volume occupied by the landfilled solid waste and cover, the efficiency of
filling and compacting operations can be adequately
judged. Additionally, the field density of the fill
can be determined to provide a better understanding
of the ability of the fill to settle.

A landfill should be designed to safely and
economically dispose of solid waste in the smallest
practical volume. These wastes come from a variety
of sources: homes, commercial establishments,
industrial plants, farms, streets, institutions and
utilities--and the differing sources require their
own methods of handling and burial. The landfill
designer should be aware of all waste types normally
disposed of from the area being served and should
call for exclusion of those materials that cannot
be safely buried. The landfill operator must be
aware of these exclusions and users of the site
notified accordingly. One of the most important
aspects of a sanitary landfill operation is the
handling of solid waste by the available landfill
equipment. This affects such diverse factors as
cost and completed use.

The wastes from residential and commercial areas and those wastes derived from industrial plant operations, exclusive of process wastes, contain a heterogeneous mixture of materials including various forms of highly compactible paper, cans, bottles, cardboard and wooden boxes, plastics, lumber, metal rods, yard clippings, food waste, rocks and soil. By themselves, boxes, plastic and glass containers, tin cans, and brush can be compressed and crushed under relatively low pressure; but in a landfill, these hollow and compressible items are incorporated within the mass of solid waste which acts as a cushion. These relatively low strength materials are prevented from deforming under the load of landfill compaction equipment. Spreading and compacting the waste in layers no greater than two feet in depth and then compacting the waste by making 2 to 5 complete coverage passes effectively reduces this problem. Brush and yard clippings should be spread and compacted near the bottom of the cell so that other more cohesive waste can be spread and compacted on top of them. If the waste is properly spread and compacted, a minimum field density (wet weight) of 800 pounds per cubic yard should be possible with solid waste received at most landfills while, as mentioned earlier, a field density of 1000 pounds per cubic yard is common.

Volume reduction through compaction at a landfill is not significant for some bulky wastes such as demolition materials and trees. Car bodies, furniture and appliances may be difficult to compact with small crawler dozers but some volume reduction can be achieved. It is best to compact these wastes with heavier equipment on ground that has not been filled.

Acceptable industrial wastes delivered to a landfill may be liquid, semiliquid, films (magnetic tape), sheets (paneling), granules, shavings, turnings, powders and other manufactured products of all shapes and sizes. The wide variety of industrial process wastes and their different chemical, physical and biological characteristics make general statements about handling difficult. It is extremely important to evaluate how these wastes can influence the environment; should an industrial waste be determined unsuitable, after discussion with the industry concerned, then the waste should be excluded. Not to be overlooked is the health and safety of landfill personnel.

When the effect of any leachate collection and treatment system has been evaluated and is determined as acceptable, the liquids and semi-solids may be

admixed with the other solid wastes or simply spread
on the existing landfill surface. They may also be
disposed of in special pits constructed in soil of
low permeability, or alternatively in areas of highly
permeable soil well above the ground water table.

Films and other light, fluffy, easily airborne
material can be a nuisance at the working face, and
they should be covered immediately when deposited
to avoid littering. Large sheets of material (plastic
or composite wood panels, etc.) can also be a nuisance;
the equipment operator should try to place or stack
these sheets parallel to each other. Random placement
of these sheets will only lead to large voids, poor
compaction, and differential settlement of the
completed landfill.

Granules, shavings, turnings and powders can be
a health hazard to the equipment operators, a nuisance
if they become airborne, and perhaps abrasive or
corrosive to landfill machine parts. Their effect
on the landfill personnel should be carefully evaluated.
Face masks, goggles or protective clothing may be
needed to prevent respiratory, eye, or skin ailments.
These wastes should be covered immediately, after
spraying with water if advisable.

Some wastes, such as paints, paint residues, dry
cleaning fluid, magnesium shavings, etc. are volatile
or flammable. These materials may be powders, solids,
or liquids and usually derive from industrial processing or commercial wastes. Depending on the degree
of flammability or how volatile they are, they may be
disposed of along with the other wastes, placed in
separate areas, or excluded from the fill. Hazardous
or dangerous materials, such as radioactive wastes or
explosives, are normally not received at landfill
sites.

The cover materials generally used at a sanitary
landfill are classified for daily, intermediate, and
final use. The most obvious identifying characteristic
of these classes is the thickness of soil used, determined by susceptibility of the soil to erosion
(wind and water) and the ability to perform identified
functional requirements. Guides for use are usually
expressed in terms of the length of time the cover is
exposed to the elements. In general, should the cover
be exposed for more than one week but less than one
year, daily cover is sufficient. Should the waste be
exposed for greater than one year, a final cover should
be used. All cover material should be well compacted.
Coarse grained soils can be compacted to 100 to 135

pounds per cubic foot,* while fine grained soils can
be compacted to 70 to 120 pounds per cubic foot.*
The important functions of daily cover are vector
control, litter control, fire break and moisture
control. Generally, a minimum thickness of 6 inches
of compacted soil is recommended to perform these
functions, at least at the end of each operating day.
At the end of the operating day, all working faces
should be covered and graded to prevent erosion and
ponding of water.

Important functions of intermediate cover, in
addition to the above, also include gas control and
perhaps service as a road base. Recommended minimum
depth is one foot with periodic grading and compacting
to insure that erosion is kept minimal and to prevent
ponding of water. Since cracks and depressions may
develop, additional soil may be needed for repair.

The final cover must support vegetative growth.
A minimum total cover depth of two feet should be
used and compacted in layers. Grading of the final
cover should be specified in the landfill design and
operations plan, and the topographic configuration
attained by careful location of the solid waste cells.
Grades should not exceed 2-4% to prevent erosion of
cover material, and the side slope should be less
than 1 vertical to 3 horizontal. Top soil from the
site should be stockpiled and reserved for placement
as final cover and should not be highly compacted.

Dust is sometimes a problem at sanitary landfills,
especially in dry climates and if the soil is fine
grained, causing excessive wear of equipment, potential
health hazards to personnel on the site, and a nuisance
to residences or businesses nearby. Dust from vehicular
traffic can be temporarily controlled by wetting down
the road with water or by using a deliquescent chemical.
Calcium chloride, applied at 0.4 to 0.8 pounds per
square yard and admixed with the top three inches of
the road surface is often effective. Frequent applications are usually required. Waste oils can also be
used as temporary dust palliatives when applied by
spraying at 0.25 to 1.0 gallons per square yard.

Salvage of usable materials from solid waste,
while a laudable concept, should be discouraged at a
sanitary landfill unless the landfill has been designed for salvaging and the appropriate processing
and storage facilities provided. Salvaging should

---

*Unit dry weight of compacted soil at optimum moisture content
for standard AASHO compactive effort.

never be practiced at the working area of a sanitary
landfill. Scavenging (sorting through waste to recover seemingly valuable items) should be strictly
prohibited at a sanitary landfill.

Satisfactory construction of a sanitary landfill
depends in part on the cooperation received from
users of the landfill, those who generate the waste,
and especially those who deliver the waste to the
working face. A pamphlet describing the purpose of
the disposal site, a definition of a sanitary landfill, wastes that can be disposed at the landfill,
hours of operation, fees charged, procedures to be
followed when disposing of waste at the landfill,
and use of the completed site should be made available.
Distribution of this pamphlet can be done through the
mail, or the public can be informed by the news media.
All vehicles entering the site for the first time
should be given a copy of the information pamphlet
to familiarize the driver and helpers with the
landfill operation.[1]

Other reference material on design and construction of the sanitary landfill include American Public
Works Association,[2] Warner, Parker and Baum,[3] Seibel[4]
and Sorg and Hickman.[5]

LANDFILL EQUIPMENT

A wide variety of equipment is on the market today
from which to select the proper type and size needed
for an efficient operation.[1-5] The most common
equipment used on sanitary landfills is the crawler
or rubber-tired tractor. A tractor can be used with
a dozer blade, trash blade, or a front-end loader.
A tractor is versatile and can normally perform all
the operations--the spreading, compacting, covering,
trenching, and even the hauling of the cover material.
The decision on whether to select a rubber-tire or a
crawler-type tractor, and a dozer blade, trash blade,
or front-end loader must be based on the existing
conditions at each individual site. Other equipment,
used normally only at large sanitary landfills, include scrapers, compactors, draglines, rippers, and
graders.

The size of the equipment is dependent primarily
on the size of the operation. Small sanitary landfills for communities of 15,000 people or less, or
sanitary landfills handling 40 tons of solid wastes
per day or less, can operate successfully with one
tractor of the 5- to 15-ton range. Heavier equipment

Table 18.1

Average Equipment Requirements[5]

Population	Daily Tonnage	No.	Type	Size (M lb)	Accessory*
to 15,000	to 40	1	Tractor crawler or rubber-tired	10 to 30	Dozer blade Front-end loader (1 to 2 yd) Trash blade
15,000 to 50,000	40 to 130	1	Tractor crawler or rubber-tired	30 to 60	Dozer blade Front-end loader (2 to 4 yd) Bullclam Trash blade
		*	Scraper Dragline Water truck		
50,000 to 100,000	130 to 260	1 to 2	Tractor crawler or rubber-tired	30 or more	Dozer blade Front-end loader (2 to 5 yd) Bullclam Trash blade
		*	Scraper Dragline Water truck		
100,000 or more	260 or more	2 or more	Tractor crawler or rubber-tired	45 or more	Dozer blade Front-end loader Bullclam Trash blade
		*	Scraper Dragline Steel wheel compactor Road grader Water truck		

* Optional--dependent on individual need.

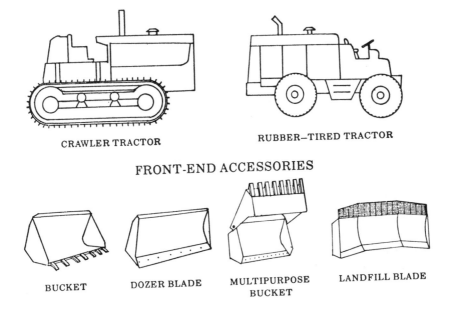

CRAWLER TRACTOR

RUBBER–TIRED TRACTOR

FRONT-END ACCESSORIES

BUCKET

DOZER BLADE

MULTIPURPOSE BUCKET

LANDFILL BLADE

Figure 18.1. Standard landfill equipment.[5]

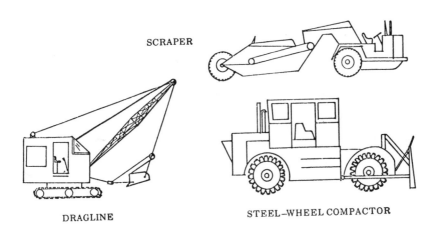

SCRAPER

DRAGLINE

STEEL–WHEEL COMPACTOR

Figure 18.2. Specialized equipment.[5]

Table 18.2

Performance Characteristics of Equipment[1]

Equipment	Solid Waste		Cover Material			
	Spreading	Compacting	Excavating	Spreading	Compacting	Hauling
Crawler dozer	E	G	E	E	G	NA
Crawler loader	G	G	E	G	G	NA
Rubber-tired dozer	E	G	F	E	G	NA
Rubber-tired loader	G	G	F	G	G	NA
Landfill compactor	E	E	P	E	E	NA
Scraper	NA	NA	G	E	NA	E
Dragline	NA	NA	E	F	NA	NA

Basis of evaluation:

1. Easily workable soil
2. Cover material haul distance greater than 1000 feet

Rating Key:

E – Excellent
G – Good
F – Fair
P – Poor
NA – Not applicable

is recommended for sanitary landfill sites serving more than 15,000 people or handling more than 40 tons per day.

Sanitary landfills servicing 50,000 people or less or handling about 115 tons or less of solid wastes normally can manage well with one piece of equipment. At these small sites, where only one piece of equipment is used, provisions must be made for standby equipment. Arrangements can normally be made with another public agency or private concern for the use or rental of replacement equipment on short notice in case of a breakdown of the regular equipment.

At large sanitary landfills serving more than 100,000 people, or handling more than 260 tons of solid wastes per day, more than one piece of equipment will be required. At these sites, specialized equipment can be utilized to increase efficiency and minimize costs. Table 18.1 offers a general guide for the selection of the type, size, and amount of equipment for various sizes of sanitary landfills.

For readers with interest in equipment used in landfill operations, very good supplemental data may be obtained from Brunner, Keller, Reid and Wheeler,[1] Seibel,[4] and Sorg and Hickman[5] in addition to commercial literature from equipment manufacturers. Table 18.2 shows average equipment requirements. Figure 18.1 illustrates standard equipment and Figure 18.2 some specialized equipment.

REFERENCES

1. Brunner, A. R., D. J. Keller, C. W. Reid, Jr., and J. Wheeler. "Sanitary Landfill Guidelines, 1970," Public Health Service Publication, Office of Solid Wastes Management Program, Environmental Protection Agency.
2. American Public Works Association. *Municipal Refuse Disposal,* 3rd edition. Interstate Printers and Publishers, (1970).
3. Warner, A. J., C. H. Parker, and B. Baum. "Plastics Solid Waste Disposal by Incineration or Landfill," study prepared for Manufacturing Chemists Association (December, 1971).
4. Seibel, J. E. "Landfill Operations," Vol. IV, Part 1, Public Health Service Publication No. 1886, Office of Solid Wastes Management Program, Environmental Protection Agency (Combustion Engineering Co.).
5. Sorg, T. J. and H. L. Hickman, Jr. "Sanitary Landfill Facts," Public Health Service Publication No. 1792, Office of Solid Wastes Management Program, Environmental Protection Agency, Second edition (1970).

CHAPTER 19

ECONOMICS AND FIELD INSTALLATIONS

Environmental conditions will determine, in part, the total cost involved in using each site. Other factors include equipment requirements, cover material needs, improvements that will be needed at the sites, and the haul time from the center of waste collection routes.

LAND

The cost of land is variable across the nation and can represent a significant portion of the total disposal cost. Land requiring a minimum of site improvements may well be in demand for other uses and consequently will bring a high purchase price. Lands requiring extensive and costly site improvements, such as wetlands, will usually not be in great demand for other uses and the purchase price will accordingly be low.

Lease and rental arrangements will keep land costs low, but will deny the landfill operator the possibility of selling reclaimed land.[1]

COVER MATERIAL

If acceptable cover material is scarce or not readily available at the working face, the costs of purchasing and hauling required soil volumes must be considered. These costs depend on the local soil conditions. Estimates of purchase costs generally range between 50 cents to a dollar per cubic yard of cover material.[1]

SITE IMPROVEMENTS

Various site improvements will have to be made
before landfilling begins. Typical improvements
include clearing and grubbing, pollution controls,
access roads, water supply, equipment housing,
fencing, scales, and sanitary facilities. Some
will be provided at all proposed sites and are not
affected by site conditions (equipment housing,
fencing, and scales). Adequate all-weather access
roads designed to handle anticipated traffic loads
from the public road system to the site will probably
be of different length. Modifications to existing
roads and attendant structures may be necessary to
provide access to the site or to shorten travel time.
The amount of clearing and grubbing will vary between
sites. The cost of water supply and sanitary services
will also vary, depending on the proximity of the
site to existing services. Other improvements such
as an earth embankment, grading and landscaping may
be required to shield the daily operation from view
and provide an appearance compatible to that of the
surrounding area.[1]

HAUL COST

Solid waste must be hauled to the disposal site
from the collection area. Ideally, the landfill
should be located at the center of the waste collection area to keep hauling costs at a minimum. This
is rarely possible, due to the lack of land and to
socio-political considerations. Solid waste is
commonly hauled by truck, the costs depending primarily on the time of travel of the collection
vehicles; a significant number of sites are more
than 10 miles from the collection area.
Strategically located transfer stations may permit
economic use of landfill sites 50 miles from the
center of collection if the additional costs can be
amortized over a sufficiently large number of tons.
Rail and barge provide economical means for hauling
large quantities of solid wastes, but savings
realized on transportation costs must be balanced
against the cost of extra handling involved. Transportation is becoming more and more a requirement in
the development of solid waste management systems if
the final disposal is to be organized on an "out-of-
town" or better still "regional" or "super-regional"
basis. The logic of this situation implies that,

specifically, long distance transport must be evaluated along with volume reduction to an ever increasing degree.

Indications from various sources, which are being further developed, show that for shipments of about 1,000 tons per day and 100 tons per car, the cost may be about $1.90 per ton to move the wastes about 50 miles. At 2,500 tons per day and the same cars, this price may drop to about $1.30 per ton for about 50 miles and to about $1.70 per ton for about 100 miles. To move 2,500 tons about 150 miles might increase the cost to about $2.05 per ton. The above tonnage data suggest that a solid waste rail-haul system requires a regional set-up to establish an economical operation.[2]

COST COMPARISON

Costs of the sanitary landfill method for refuse disposal cover a wide range. The total cost includes the land cost plus site development plus operating and equipment costs. Since the land should increase in value, even in a remote area, the cost of the land from a long range point of view is sometimes neglected.[1]

Recent discussions indicate an almost unanimous agreement that costs will usually range from $0.75 to $1.50 per ton. A cursory evaluation of landfill costs is derived from the following equation (not including transportation costs):[4]

$$\$/Ton = 0.50 + \frac{6,000}{tons\ per\ year}.$$

For very large landfill operations (over 200 tons per day), it will be apparent that landfill costs lower than $0.60 per ton should be achievable.

For fiscal 1968-1969, in the city of Los Angeles, indicated landfill disposal costs varied from $0.30 per ton to $0.60 per ton, excluding land costs, within the city and $1.20 per ton for disposal in outlying areas. The overall cost, excluding land, was $0.67 per ton. The city of Los Angeles in this same period was operating a 4,000 tons per day land-fill with a working force of 18. It should be noted that no credit is taken for the land after use, which could conceivably be more valuable after landfill has been completed than the original cost.

Stone[3] gives interesting data in terms of 1961 dollars on a comparison of "normal" versus "scientific" landfill. Scientific fill construction entails the

addition of water to facilitate compaction plus some
extra compaction. No water is used in the normal
landfill method.

The exclusion of soil from a landfill should
eliminate the need to spend about $0.25 per cubic
yard of earth for excavation, hauling and compaction
and make available additional fill for refuse disposal. Similarly, the addition of water, better
compaction and the elevation of fill depth should
increase the in-place density of the refuse by 50%.
If the cost of water added to a landfill is estimated
to be worth $40.00 per acre foot, then with the
admixture of approximately 84 gallons per cubic yard
of trucked refuse, the cost is approximately $0.01
per cubic yard. If it is also assumed that landfill
volume capacity is worth $0.05 per cubic yard, then
a cost comparison can be made between a scientific
and a normal (Los Angeles) sanitary landfill. Obviously each particular landfill location requires
separate evaluation. The scientific method showed
a $2,100 saving in the cost for soil placed in the
fill and a $1,600 saving in land used. Additional
labor and materials were estimated to cost $2,000
plus a cost for water of $420, leaving a net saving
of $1,280. This amounted to 6% of the total costs,
a reduction of 44% on the earth needed, and 34% on
the landfill volume required.[3]

Sorg and Hickman[4] consider that to compare the
true cost of sanitary landfilling with that of incineration or composting, it is essential that the
costs and returns of the initial investments and
the hauling costs be considered along with the total
disposal costs including the disposal of incinerator
residue and noncompostable materials. The hauling
costs of a collection system that uses the sanitary
landfill disposal method may be higher than the
hauling costs of a system using incineration or
composting, since sanitary landfills are generally
located farther from the waste-generating area than
are incinerators or compost plants. A sanitary
landfill, however, may increase the value of a plot
of unusable land by converting the site to a playground, golf course, or park, thereby obtaining a
major investment cost advantage over incineration
and composting. Sanitary landfill operating costs
are illustrated in Figure 19.1.

LOS ANGELES COUNTY SITES

In a paper on the techno-economics of landfilling,
Parkhurst[5] discusses the current practice of landfill

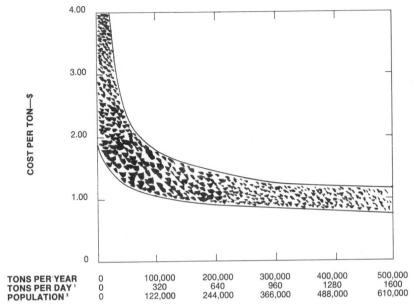

TONS PER YEAR	0	100,000	200,000	300,000	400,000	500,000
TONS PER DAY [1]	0	320	640	960	1280	1600
POPULATION [2]	0	122,000	244,000	366,000	488,000	610,000

[1] Based on 6-day work week.
[2] Based on national average of 4.5 lbs per person per calendar day.

Figure 19.1   Sanitary landfill operating costs. [4]

in California. Even though the metropolitan area of
Los Angeles County is probably as badly afflicted
with urban sprawl as any similar region in the country,
virtually all of the solid waste in Los Angeles is
presently being disposed in sanitary landfills. There
are 25 sites presently operating in the county which
collectively dispose of approximately 10 million tons
of solid waste per year. With a tributary population
of approximately 7 million, this is equivalent to
nearly 8 pounds per person per day. Data are shown
in Table 19.1.

Numerous engineering and planning studies have
separately and repeatedly arrived at the conclusion
that the only reliable long-range solution for solid
waste disposal in Los Angeles County lies in land
disposal. Countywide projections by the sanitation
districts indicate that for a 50-year period,
approximately 1.4 billion cubic yards of space will
be needed. At first glance it would appear that
this might result in covering the entire county with
refuse, but this is far from correct. The sanitation
districts place great emphasis upon using only those

Table 19.1

Existing Landfills*--*Los Angeles County*[5]

No.	Site	Sanitation Dist.	Municipal	Private
		Fill Rate (Tons/Year)		
1	Palos Verdes Landfill	1,600,000		
2	Spadra Landfill	180,000		
3	Mission Canyon Landfill	1,400,000		
4	Scholl Canyon Landfill	500,000		
5	Calabasas Landfill	260,000		
6	Puente Hills Landfill	940,000		
7	Sheldon-Arletta Pit		510,000	
8	Toyon Canyon Landfill		730,000	
9	Burbank City Landfill		120,000	
10	City of Whittier Landfill		36,000	
11	North Valley Refuse Center			360,000
12	Bradley Avenue Dump			600,000
13	Hewitt Pit			280,000
14	Tujunga Pit			370,000
15	Azusa Rock and Sand			250,000
16	Owl Park			84,000
17	Canyon Park Development			95,000
18	B.K.K. Co.			560,000
19	Operating Industries			900,000
20	Norwalk Dump Co.			14,000
21	Kobra			41,000
22	Ascon			300,000
23	Lancaster Dump			18,000
24	Antelope Valley Landfill			20,000
25	Chiquita Canyon Landfill			28,000
	Subtotals	4,880,000	1,396,000	3,910,000
	Grand Total for All Sites		10,186,000	

*Estimated existing landfills in Los Angeles County as of June 1970.

lands for refuse disposal that can accommodate large amounts of waste for each acre of land used. Land that cannot accommodate at least 100,000 cubic yards per acre is given little consideration, and some of the sites that are being acquired yield twice this amount.

With this guideline on land utilization, Parkhurst estimated that the entire 50-year need for Los Angeles County can be met on less than 10,000 acres of land. Completed landfill sites undergo substantial settlement and are necessarily limited as to the types of development which can be placed on them.

Figure 19.2 shows the total operating and maintenance costs on a per ton basis that were experienced at six sites in Los Angeles County during the 1970 calendar year. The operating volume at these facilities ranged from 600 to 5,000 tons per day. Costs range from approximately $1.10 per ton at the smallest site to $0.70 per ton at Palos Verdes, one of the largest landfills operating in the country. These costs

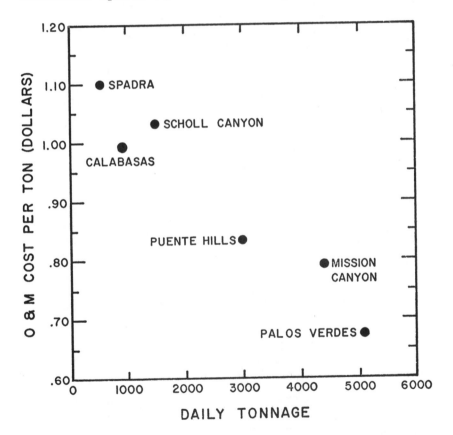

Figure 19.2. *Operating cost at sanitation district landfills--1970.*[5] *Note: Operating costs include labor, overhead, amortization and all other costs as described on the following page.*

include labor, amortization and maintenance of equipment, materials and supplies, administrative overhead and supervision, and all other expenses involved in the operation. When the cost of land and major capital improvements are included, the total cost for landfilling would rarely exceed $1.50 per ton. Since typical residential curbside service including disposal by landfilling will average about $15.00 per ton, it is evident that the cost for landfilling is only about 10% of the total cost of refuse collection and disposal. The costs shown are representative of operations where the refuse is placed and compacted with crawler tractors and covered with earth using conventional earth moving equipment. Currently, there is much discussion in the literature favoring other methods such as shredding, baling, or use of a high-pressure extrusion mold. Any one of these processes could be expected to cost an additional $2.00 per ton.

Parkhurst comments that in many parts of the country, ample sites are available which if set aside and reserved now for this purpose could provide for the disposal of solid wastes for many years. Almost universally the deterrent to proceeding with such programs is the lack of financing. Most local governmental jurisdictions are so overwhelmed with the financial problems of immediate concern, such as welfare and other social problems, that it is difficult for them to consider investment of their limited funds in a waste problem that will develop some 10 to 20 years hence. A significant portion of today's solid waste disposal industry is conducted through private enterprise. When it comes to investing in land sufficient to provide for half a century of waste disposal, private enterprise cannot be counted upon to do the job. If for no other reasons, property taxes on the land and cost of money will almost always negate the wisdom of such investment.

If the federal or state governments, or both, would set up a low interest loan plan, local public agencies could then borrow the necessary funds to acquire the needed sites and repay the loans from revenues. In short, the prime need is money to invest now in solutions which are known to work today and will continue to work in the future. Federal or state funding would be best spent in developing practical solutions for the waste disposal problem.[5]

OTHER U.S. SITES

Some sanitary landfill data from other parts of the United States are shown in the following:

(a) The Des Moines metropolitan area operates two sanitary landfill sites available for anyone to use at a fee of approximately $1.10 per ton.

(b) For Santa Clara, California, it is stated that a net cost of $1.50 per ton is involved as follows:

Population served	100,000
Refuse production	5 lb/day/person
Density of refuse	1000 lb/cu yd
Refuse filled	91,000 tons per year
Landfill volume (including earth and intermediate cover)	136 acre-feet/year
Landfill depth (including cover)	12 feet
Area used	113 acres per year
Disposal cost	$114,000 per year
Land cost	$2,000 per acre or $23,000 per year
Overall cost	$137,000 per year
Net cost per ton	$1.50

It should be noted that no credit is taken for the land after use, which could conceivably be more valuable after landfill has been completed than the original cost.

(c) The Meadowlands in Hackensack, New Jersey is probably the largest landfill operation in the world. Located only some fifteen minutes from New York City, it presently caters to the wastes from 118 municipalities as well as an estimated 1.4 million tons per year from the State of New York. The site is 13,000 acres in extent and contains 11 major facilities handling some 15 million tons per year. Such a site is obviously favorable to a massive engineering approach to a number of advanced concepts in solid wastes disposal. It is hoped that the proposed Development Commission for the Meadowlands, formed to develop this land, will seize this opportunity to demonstrate how the municipal wastes of a large population center can be most effectively handled to the best interests of the community.

(d) It has also been estimated that for 13 cities in Pennsylvania, whose future population might total in excess of 10,000,000 persons, suitable land is available within twenty miles to accommodate the needs for at least one hundred years.

The New York State Department of Environmental Conservation, Bureau of Solid Waste Engineering in a 1971 booklet entitled "Sanitary Landfill," comments on the use of a completed landfill:[6]

The landfill plan should provide for a specific use of the area upon completion. Rather than requiring expensive excavation and regrading, the necessary contours should be established in advance. For example, a golf course can be constructed on rolling terrain while a parking lot would require a flat graded surface.

Permanent buildings should not be constructed over fill areas because of settling problems and underground gas production which could enter sewers or basements and develop explosive conditions. With special foundation structures and provisions to vent gas, it is possible to build on fill areas; however, it is much cheaper if an advance plan is drawn up leaving unfilled areas for building construction.

It is recommended that large sites be planned to allow multiple use. For example, while waste is being landfilled in one portion of a site, other areas could continue to be used for farming. The entire site need not be filled before the planned final use is commenced. Portions of the site can be developed for final use as soon as filling is completed. The multiple use concept allows for maximum utilization of land and also builds good community relations. With proper planning, airports, swimming pools, ski areas or practically any other type of facility can be made from a completed landfill. Advance planning is essential, however.

Some sanitary landfill costs in New York State are shown in Figure 19.3.

In a discussion of various waste disposal data, APWA[7] states that the parameters for solid wastes are known only partially, and at best only on a sporadic basis for an extremely small number of the communities in the nation. Most communities do not collect all the solid wastes produced within their boundaries because of either voluntary or mandatory self-disposal by the producers themselves. Furthermore, the amount of refuse produced and collected varies according to the nature of the community's

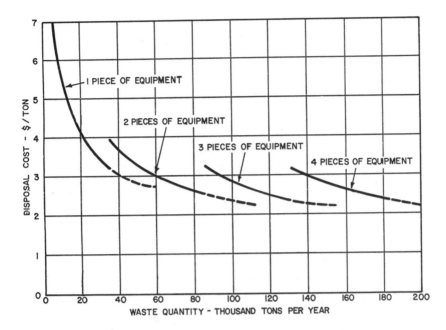

Figure 19.3.   *Sanitary landfill costs--New York State.*[6]
*Note: The dashed portions of the curve indicate
overtime or second shifts allowing the site to
be operated without purchasing additional equipment. Graph was developed by Malcolm Pirnie,
Inc.*

economic function (industrial, commercial, residential),
according to the habits and styles of living of the
residents, income levels, climate, season of the year,
and the frequency of collection.

The collection of solid wastes involves storage
at the place of origin and transportation to the
point of disposal. The method of collection of refuse
is related to the method of disposal. Separate garbage
collection of refuse is essential if hog feeding of
cooked garbage or other salvage collections are included in the methods of disposal. Combined collection
of mixed refuse including garbage is made if the
disposal is done by sanitary landfill or incinerator.
Combined collection, of course, permits combined
storage.

The disposal operations in a community are handled
in various ways. Excluding disposal by the producers
themselves, public agencies sometimes perform the
complete task with public employees; in other cases

public agencies contract with one or more private,
profit-making organizations; in still other instances,
all waste disposal is achieved through agreements
between the individual producer and private enterprise
or through various combinations of the above three
methods. Data on the public-private relationship in
the ownership of disposal facilities are much more
difficult to obtain. The 1965 APWA survey on collection practices indicates that 10.3% of the contractors
and 13.4% of the private collection organizations do
not use disposal facilities operated or provided for
by public agencies.

A limited 1966 APWA survey[8] indicates that about
15% of the solid wastes in communities of more than
10,000 population are disposed of in open dumps,
about 65% in sanitary landfills, about 18% through
incineration, and about 2% by other methods. Field
surveys in some regions indicate that only about 10%
of the sanitary fills apply cover material on a daily
basis. Thus, about 90% of the sanitary landfills
might actually be classified as open dumps, including
modified landfills or modified open dumps.

Most open dumps are found in the small rural areas
having a population of less than 2,500 people. Rural
areas are estimated to account for 12,250 to 15,000
of the open dumps and for only 125 to 150 of the
sanitary landfills. Urban areas are estimated to
account for 5,200 to 6,300 open dumps and 850 to
1,000 sanitary landfills. The estimated 1965 distribution of the number of open dumps and sanitary
landfills is given in Table 19.2. Again, a margin
of error of 10% either way should be allowed.

In evaluating surveys on the use of open dumps,
modified landfills, and sanitary landfills, it is
found that the inventories made by private consultants
account for a much higher share of open dumps in the
total disposal capacity than those shown on question-
naire surveys. The ages of open dumps and sanitary
landfills are difficult to estimate. The service
life of these installations varies greatly, depending
mainly upon fill-depth, degree of compaction, and
size of the area.

Non-captive refuse collection and disposal
facilities are generally owned by local governments
or private profit-making organizations or individuals.
None are known to be owned by state governments, state
agencies, the federal government, or by private,
non-profit organizations and cooperatives. The
ownership relationship between local governments
(municipalities, townships, counties and special

Table 19.2

Estimated Distribution of the Number of Open Dumps and Sanitary Landfills By Community Size, 1965[8]

Community Population	No. of Places in U.S. 1960	Avg. No. of Open Dumps and/or Sanitary Landfills per Community	Ratio of Open Dumps to Sanitary Landfills (Percent)	Distribution by Community Size			
				Number		Percent	
				Open Dumps	San. Fills	Open Dumps	San. Fills
URBAN (1000's)							
1,000 or over	5	4.0	25:75	5	15	0.0	1.4
500–999.9	16	3.5	34:66	19	37	0.1	3.5
250–499.9	30	3.0	50:50	45	45	0.2	4.2
100–249.9	81	2.5	60:40	120	82	0.6	7.7
50– 99.9	201	2.0	70:30	281	121	1.4	11.2
25– 49.9	432	1.5	75:25	485	163	2.5	15.2
10– 24.9	1,134	1.0	80:20	910	224	4.7	21.0
5– 9.9	1,394	1.0	90:10	1,255	139	6.6	13.0
2.5– 4.9	2,152	1.0	95:05	2,050	102	10.6	9.4
Under 2.5*	596	1.0	99:01	590	6	3.1	0.6
(Subtotal Urban)	(6,041)	–	(86:14)	(5,760)	(934)	(29.8)	(87.2)
RURAL							
1,000–2,500	4,151	1.0	99:01	4,110	41	21.2	3.8
Under 1,000	9,598	1.0	99:01	9,502	96	49.0	9.0
TOTAL	19,790	–	94:06	19,372	1,071	100.0	100.0

*Communities with less than 2,500 population located in urbanized areas including the Standard Metropolitan Statistical Areas.

districts) and proprietary profit-making organizations
is estimated to be as follows:[8]

In the absence of any data on the current value
of refuse disposal investments, APWA decided to use
replacement values taking into account the past conditions in the state-of-the-art of the disposal
equipment technology. Local governments do not tend
to use capital investment accounting methods including
depreciation and other value (land value) adjustments.
Sanitary landfills, if properly operated and completed,
often represent land investments of considerable value.

Taking the average size of a sanitary landfill at
15 to 30 acres with a price per acre of $1,500, APWA
calculates that the past investment value for such an
average landfill ranges from $50,000 to $112,000 on
the assumption that land costs are 40 to 45% of the
total. For open dumps, land cost alone is used in
the calculations, and considering the large number
of open dumps in small rural communities and the
smaller sizes of such dumps, an investment of $500
for each is estimated. Obviously land prices in
urban areas, where most sanitary landfills are sited,
are substantially higher. Excluded is the investment
value of completed sanitary landfill areas which are
still held as property of the community. In case of
a sanitary landfill, the land value usually increases
because of land improvements produced by proper
sanitary landfill operations. Thus in a strict
sense, the current value of a sanitary landfill cannot
be compared to the current value of an incinerator
installation.

The development "construction" cost for sanitary
landfills includes access roads, water, drainage
facilities, equipment sheds, fencing, lighting and
site beautification. A 1964 survey of ten sites in
Pennsylvania indicates that these costs are approximately $55,000 for a 30-acre site, varying considerably
according to terrain, location and size, and excluding
the acquisition cost for land and operating equipment.
Converted to a 10-acre site, the costs are estimated
at about $18,000. The equipment costs are estimated
to average $35,000 to $40,000 per site.

User charges cover annual maintenance and operation expenses plus debt services. According to APWA
surveys,[8] it is estimated that about 35 to 36% of
the communities finance their refuse collection and
disposal operation through service charges, 50 to 52%
through general taxes, and 12 to 15% through a combination of taxes and service charges. The data show
that the smaller communities tend to rely more on
service charges than do the larger communities.

Industrial and technological changes plus an increase in living standards result in the production of ever-increasing quantities of refuse. The estimated capital requirements for non-captive refuse collection and disposal facilities during the 1966-1975 decade are estimated to be at least $2.42 billion in 1965 dollars. This estimate is based on a survey of the capital investment needs for waste disposal facilities recently conducted by the APWA in 47 communities and the findings of the previous analyses. The amounts of these capital investment demands are estimated to be as follows: $1.420 billion for collection equipment, storage and maintenance facilities, $340 million for sanitary landfills including land and equipment, and $660 million for incinerators. The sources of financing for these capital outlays include appropriations from tax sources, tax exempt municipal bonds, borrowings from banks, and private venture capital.

In estimating the sanitary landfill capital investment needs, it is assumed that open dumps will be eliminated wherever feasible. However, inert waste materials, such as incinerator ash and certain demolition wastes, do not require sanitary landfills for adequate disposition. Furthermore, open dumps tend to be smaller in area than sanitary landfills. Thus a number of open dumps will not be converted to sanitary landfills but will be used for the disposal of inert materials. Consequently it is assumed that about 30% of the existing open dumps in urban areas will not be converted to sanitary landfills, leaving about 4,000 open dumps to be converted. Since not all of these open dumps are located in or near metropolitan areas, nor do they belong to outlying communities in a metropolitan complex, their capital investment values for conversion to landfills is calculated at $50,000 each. In turn, the conversion of open dumps to sanitary landfills is estimated to require about $200 million during the 1966-1975 decade.[8] In addition, it is estimated that about half of the existing 1,000 sanitary landfills will need replacement in the 1966-1975 decade at an average cost of $80,000 each. This will add $40 million to the sanitary landfill investment needs.

It seems that the waste disposal needs in rural areas do not presently justify the idea that each of the 13,600 communities be required to operate a sanitary landfill. Consequently, it may be assumed that refuse disposal in rural areas will be operated, more or less, as part of a country-wide system. This

in turn might suggest that about 70% of the existing
open dumps in rural areas will be closed. Because
of their smaller size and lower land costs, the
capital investment needs for sanitary landfills in
rural areas are estimated at $25,000 each, including
a part of the cost for the required equipment. Thus,
the capital investment needs for 4,100 sanitary
landfills in rural areas are estimated to be approximately $100 million. The equipment for sanitary
landfills in rural areas will not be used on a full-
time basis for landfill operations.

The foregoing estimates appear to be reasonable
in the light of the capital requirement projections
made by 20 metropolitan or regional planning commissions in urban and in some urban/rural areas. These
agencies in 1966 estimated that each of them should,
realistically, spend an average of $7.5 million
during the 1966-1975 decade on capital investments
for refuse collection and disposal facilities.
Since there are 216 metropolitan urban areas in
this country, their total capital investment needs
are calculated at $1.6 billion. Since such areas,
however, account for about 70 to 75% of the population, the total U.S. investment needs on this basis
can be extrapolated to $2.1 billion to $2.3 billion.

If the projected needs were to be financed over
the next decade in equal proportions, the annual
investment would amount to approximately $240 million
per year. If the backlog, which is estimated to be
at least 34% of the investment needs, were to be
funded during the first year of the decade, about
$820 million would be required. Spreading the re-
maining $1.6 billion evenly over the ten-year period
would add $160 million to the first year's require-
ments. Thus, it would be necessary to provide more
than 40% or $980 million of the total $2.42 billion
capital investment needs during the first year. The
remaining $1.44 billion would be required at the
rate of $160 million annually during each of the
remaining nine years. Of this amount, approximately
$90 million would be required by local government,
and $70 million by private entrepreneurs.

According to the study, $428 million, or 17.8%
of the total investment needs are estimated to be
needed in rural communities. In addition, it is
estimated that communities in urban areas with a
population of less than 2,500 people will require
about 2.2% of the total capital need. It is also
noted that most of these communities receive their
refuse disposal service in conjunction with that of

other urban communities in metropolitan areas.
Therefore, it is estimated that agricultural areas
and communities with a population under 2,500 persons will require 20%, or $488 million, of the total
capital investment needs. In turn, 80% or $1.932
billion would be spent in communities with a population of 2,500 or more people. According to the
U.S. Census, people living in such communities are
considered as living in urban areas.[8]

REFERENCES

1. Brunner, O. R., D. J. Keller, C. W. Reid, and J. Wheeler.
   "Sanitary, Landfill Guidelines--1970," review draft of
   Public Health Service publication, Office of Solid Wastes
   Management Program, Environmental Protection Agency.
2. Warner, A. J., C. H. Parker, and B. Baum. "Plastics
   Solid Waste Disposal by Incineration or Landfill," study
   prepared for Manufacturing Chemists Association (December,
   1971).
3. Stone, Ralph. "Scientific Analysis of Sanitary Landfills,"
   APWA Yearbook, pp. 250-260, (1961).
4. Sorg, T. J. and H. L. Hickman. "Sanitary Landfill Facts,"
   Public Health Service Publication No. 1792, Office of
   Solid Wastes Management Program, Environmental Protection
   Agency, Second edition (1970).
5. Parkhurst, J. D. "Techno-Economics of Landfilling,"
   paper prepared for National Conference on Solid Wastes
   Disposal Sites, APWA, pp. 9-20 (1971).
6. New York State Department of Environmental Conservation.
   Bureau of Solid Waste Engineering, "Sanitary Landfill,"
   (1971).
7. American Public Works Association. *Municipal Refuse
   Disposal*, 3rd edition. Interstate Printers and Publishers
   (1970).
8. American Public Works Association. "Solid Wastes--The
   Job Ahead," highlights of a report by APWA Research
   Foundation, *APWA Reporter*, pp. 5-11, 25 (August, 1966).

CHAPTER 20

ADMINISTRATION

THE COMPLETED SANITARY LANDFILL

Reclaiming land by filling and raising the ground surface elevation is one of the greatest benefits of sanitary landfilling. The completed landfill can be used for many purposes, all of which require planning before actual landfilling begins. The decomposing solid waste imparts characteristics to the fill peculiar only to sanitary landfills, and because of these characteristics, periodic maintenance of the completed landfill is required.

Characteristics

The landfill designer should be fully aware of the characteristics of a completed landfill before he begins the design so that he can include gas and water controls, cell configuration, and cover material specifications for the planned use. Unlike an earth-fill, which consists primarily of dirt, a sanitary landfill is constructed of a number of cells consisting of a great variety of materials with different physical, chemical and biological properties.[1,2]

Decomposition

Most of the materials in a sanitary landfill will decompose but at different rates and with different results. Food waste is readily decomposed, moderately compactible, and forms organic acids that further aid decomposition. Garden wastes are resilient and difficult to compact, but generally they are rapidly decomposed. Paper products and wood decay at a much slower rate than food waste. Paper, especially when

wet, is easily compacted and pushed into voids, while
lumber, tree branches, and stumps are difficult to
compact and hinder compaction of adjacent waste.
Car bodies, metal containers, and household appliances
can be compacted and slowly rust within the fill;
organic acids from decomposing food waste contribute
to the degradation of these metallic wastes. Glass
and ceramic are usually easily compacted but do not
degrade. Plastics are resilient but still compactible.
Rubber decomposes very slowly and most plastics decompose at a negligibly slow rate; however, cellulosics
biodegrade at a significant rate. Leather and textiles
are slight resilient, but can be compacted; they decompose at a much slower rate than garden and food
wastes. Rocks, dirt, ashes, and construction rubble
do not decompose and can be easily worked and compacted into the solid waste mass, except for boulders
and large pieces of broken concrete and asphalt.[1]

Density

The density of solid waste in a landfill is quite
variable. A well constructed landfill can have in-
place densities (wet weight) of solid waste as great
as 1500 pounds per cubic yard, while a poorly compacted solid waste may have a density of only 500
pounds per cubic yard. Generally, 800 to 1000 pounds
per cubic yard can be achieved with moderate compactive effort. Soft and hard spots can occur within
the fill as a result of different decomposition rates
and compaction of the waste. The density of the
waste influences other characteristics of the landfill
such as settlement and bearing capacity.

Settling

A particularly important characteristic of a
sanitary landfill is the settling resulting from
waste decomposition, filtering of fines, superimposed loads and its own weight. Bridging of solid
waste during construction of the fill results in
voids that are filled or collapse when the waste
stabilizes. Fine particles from the cover material
and overlying solid waste sift into voids. Decomposition of waste reduces its bridging strength and
causes the material to collapse. The weight of
overlying solid waste and cover material consolidates
the fill. Superimposed loads from addition of more

cover material or location of a building on the fill also causes consolidation.

The most significant cause of settlement is waste decomposition, which is greatly influenced by the amount of water in the fill. A landfill will settle at a slower rate when it is anaerobic and has limited water available for chemical and biological waste decomposition than when adequate water is available.

Anticipated settlement depends on the type of waste disposed, volume of cover used with respect to the volume of waste disposed, and the compaction achieved during construction. A fill composed only of construction and demolition debris will not settle as much as fill constructed of residential solid waste. A landfill constructed of highly compacted waste will not settle as much as a poorly compacted waste. Several problems are associated with settlement of a sanitary landfill. Wide cracks can develop in the cover material exposing waste to rats and flies, allowing water infiltration and permitting gas to escape. The topography of the surface can become hummocky, causing rainfall to pond and infiltrate the fill. Structures on the landfill may distort and possibly fail. Because all landfills inevitably settle, the landfill surface should be periodically inspected and soil added and graded when necessary.[1]

Bearing Capacity

Another important characteristic of a completed sanitary landfill is bearing capacity, a measure of the ability of the landfill to withstand foundation loads without shearing. Data on the bearing capacity of a landfill is very limited but appears to be in the range of 500 to 800 pounds per square foot. Greater density of solid waste in the landfill provides greater shearing resistance and hence larger bearing capacities. It is important to recognize that there is no definite procedure for interpreting the results of bearing tests on solid waste and, therefore, any value obtained should be viewed with extreme caution.

It will be very apparent that many of the problems of landfill construction, operation and planning for future use are associated with the variable nature of the wastes deposited at the site for disposal. It is for this reason that many authorities are giving increasing attention to the advantages of a precommunution process prior to the actual landfill. Such a

precomminution will increase the density of the
wastes, make compaction easier and less costly on
equipment, promote more rapid decomposition, reduce
the possible formation of gas pockets and return
the completed site to use at an earlier date. When
the volume is sufficiently large, precomminution
could be combined with a reclaiming operation whereby
certain valuable materials, such as metals and perhaps glass, could be recovered. In addition to whatever values might result, the savings in volume of
landfill are not insignificant and would extend the
use-life of the site.

It is also to be considered that precomminution
at earlier transfer stations could reduce the number
of trucks or railroad cars needed to transport the
wastes to the landfill site. While precomminution
is being practiced in other forms of disposal such
as incineration, pyrolysis and composting, its regular
use in any sanitary landfill is encountered much less
frequently.[1,2]

Uses

There are many uses to which a completed sanitary
landfill can be put. Some of these include green
areas, recreation, agriculture and construction of
buildings.

Green Area

Use of the completed sanitary landfill as a green
area is most common. Expensive structures are avoided,
and an area free of the busy day-to-day life is established for the pleasure of the community. Some work
is required, however, to maintain the quality of the
site environment. Erosion of the fill surface by
wind and water should be prevented. The cover material
should also be graded to prevent ponding of water
which could infiltrate the fill. Gas and water monitoring stations, installed during construction, should
be periodically sampled until the landfill stabilizes.
Gas and water controls and drainage from the landfill
surface will also require periodic inspection and
maintenance.[2]

Where the layer of final cover is thin, vegetation
on the landfill surface should be limited to shallow
rooted grass, flowers and shrubs, since decomposing
solid waste may be toxic to those plants whose roots
penetrate through the bottom of the soil.

Agriculture

A completed sanitary landfill can be made produc-
tive by turning it into pasture or crop land. Accumulation of landfill gas in the overlying soil or
use of deep-rooted crops should be avoided.

Recreation

Completed landfills are often used for recreational
facilities. Specific applications include ski slopes,
toboggan runs, coasting hills, ball fields, golf
courses, amphitheaters, and parks. Some small light
buildings, such as concession stands, sanitary facilities and equipment storage, are usually required at
recreational areas. These should be constructed to
keep settlement and gas problems at a minimum. Usual
problems encountered are similar to green areas:
ponding, cracking and erosion of cover material.
Periodic maintenance usually includes regrading,
reseeding, and replenishing cover material.[1]

ADMINISTRATION

The size and scale of operation of the sanitary
landfill will influence the mechanics of its administration. Complex operations will necessitate more
intricate administrative procedures than will smaller,
simpler landfills. The purpose of administration,
though, remains the same--to consolidate and coor-
dinate all resources necessary for the sanitary
disposal of solid waste by employing these resources
in the most efficient manner.[1,2,4]
Normally, the responsibility for operation of
the sanitary landfill will be determined by the
existing community administrative agency structure.
Each community will have to consider its own particular circumstances before establishing the
administration for the landfill.

Municipal Operations

In most municipal operations the administrative
responsibility will be assigned to the department of
public works. Usually one division under this department will be in charge of the community's solid waste
management. As the scope of this division's activities

increases, it becomes desirable to further subdivide
the division into functional sections. A possible
arrangement would have two sections, one responsible
for collection and the other for disposal. Regardless of organizational structure, collection and
disposal plans and operations must be coordinated
to achieve satisfactory and economical management
of the communities' solid waste.[1,3]

Special Districts

Many states have enabling legislation which permits formation of special purpose districts. In
some cases these can be created for solid waste
disposal. Generally, such legislation includes
provisions for a tax levy to provide funds for the
operation. These districts are advantageous in
that they can serve many political jurisdictions.
Before any special district is considered, the state
laws applicable to these districts should be
investigated.

Administrative Functions

The functions of the administrative agency are
very broad. It is responsible for proper solid waste
disposal which includes planning, design, financing,
cost accounting, operation, personnel, public relations, and establishment of minimum standards for
proper solid waste disposal.[1]

Finance

Sanitary landfill capital costs include land,
equipment, gas and water control devices and site
improvements, while operating costs include wages,
salaries, utilities, fuel, and equipment maintenance.
There are several sources of funds to meet these
costs.*

----

*Additional information is available from *Solid Waste Manage-
ment: Financing*, one of a series of guides developed by the
National Association of Counties Research Foundation.

General Fund

The general fund, derived from taxes, is generally not large enough to provide the capital costs for a sanitary landfill. These funds, however, are often used to provide the operating costs of a landfill.

General Obligation Bonds

The use of general obligation bonds is often a favorable method of financing the capital costs of a sanitary landfill as these bonds have a low interest rate and high marketability. These bonds are secured by the pledge of real estate taxes, and, in effect, the real estate within the tax district serves as security for borrowed funds. Usually, state statutes limit the amount of debt a community can incur. If the community has a substantial debt, this method may not be practical. In some cases these bonds are retired with revenues generated by the landfill operation, thus minimizing *ad valorem* taxes necessary for bond retirement.

Revenue Bonds

Revenue bonds differ from general obligation bonds in that they are secured by the ability of the project to earn enough revenue to pay the principal and interest of the bond issue. In this case, fees are charged to landfill users in amounts necessary to cover all capital and operating expenses. This method of financing a municipal sanitary landfill is the most desirable as it forces the administering agency to follow good cost accounting procedures and allows the agency to be the sole beneficiary of cost saving procedures. Also, the producer or generator of solid waste is forced to pay the true cost for its disposal.

Fees

User fees are primarily a source of operating revenue, though a municipality can employ them to generate funds for future capital expenditures. Although fees necessitate more work and expense by requiring weighing, billing and collecting, this provides insight into the management and operation of the landfill. The fees can be adjusted to cover not only the operating and capital cost of present landfills but also to provide a surplus for future land and equipment acquisition. They are similar to revenue bonds but do not provide the capital outlay needed to start a sanitary landfill.

Usually commercial haulers are billed on a per ton basis while individual homeowners are billed on a per load basis. As fee operations require that collection vehicles be recorded at the gate, this provides an additional control on what wastes are being received at the landfill.

Cost and Operations Control

When the sanitary landfill is in operation, the administration should monitor and control the cost of operation. One tool for operational control is cost accounting which isolates costs of ownership and operation and matches these costs against revenues. The important costs of operation include:

Wages and salaries
Maintenance and fuel for equipment
Utilities
Depreciation and interest on buildings and equipment
Overhead

Such a system of control should measure the amount of waste disposed at the fill, either in tons or cubic yards, to provide a relative measure of efficiency. A cost accounting system recommended for use at a sanitary landfill has been developed by the Office of Solid Waste Management Programs (OSWMP).[1,3]

Performance Evaluation

In most cases there will be a health agency at the state or county level which determines if a sanitary operation is being conducted properly. To insure this, the administrative agency should conduct its own performance evaluation through someone at the administrative level rather than the operating or supervisory level. The administrative performance evaluation should be at least as stringent as those of the health agency. While operating and super-visory personnel should know that these inspections will occur with a specified frequency, such as once per month, they should not know on what day they will occur. This will help insure a more representative inspection.

Personnel

The proper operation of a sanitary landfill depends upon competent employees. To secure and retain these employees, the administration must have a systematic plan for personnel management. First, the administration must define the jobs which must be performed at the sanitary landfill. One method of isolating the necessary tasks is by the nature of the work.

Once the job areas are defined, management must determine how many employees are needed to accomplish the necessary tasks. Except in the largest operations, the administrative functions will be handled by a staff which serves many different operations such as collection, transfer, incineration, and landfill. After the jobs are defined, management must hire the necessary employees. Governmental operations normally will have a civil service system describing the hiring and career advancement procedures and outlining the job descriptions. Private operations have more latitude in their employment practices. Once an employee is hired, management must see that he is trained properly.

All employees should be grounded in the proper safety and emergency procedures. Personnel relations after hiring and training are critical to any operation. Wages at the landfill must be comparable with similar employment elsewhere.

Public Relations

Public relations is one of the most important administrative functions. Solid waste disposal sites represent an extremely emotional issue, particularly to those who live in the vicinity of the site. Citizens often associate a proposed sanitary landfill with the open dump.[1,3,4] As with any other civic improvement, administration must convince the public of the advantages of a sanitary landfill. This is a tedious process but can be accomplished by explanation and education. Examples of properly operated sanitary landfills in proximity to residential areas should be shown. Benefits to be derived by the completed use

of the site, such as a park or playground, should
be emphasized.

Once a site is established, management must insure proper operation. Possible items which are
highly visible to the public are litter, dust, and
vectors. Management should periodically inspect
the access roads and site to prevent problems in
these areas.

A key aspect of public relations is the procedure
for handling any citizen complaints. Reported de-
ficiencies in operating methods or employee courtesy
should be investigated and acted upon promptly. If
this practice is followed, citizens will be less
hostile towards the operation, and employees will
be more conscious of their performance. A sanitary
landfill represents a positive and relatively inexpensive step communities can take toward providing
a safe and attractive environment.

The specific role of plastics in sanitary landfill operations is covered in detail in another
chapter and it suffices here to state that, in the
opinion of experts, no additional problems due to
the presence of plastics appear to be involved.

TRENDS

The number of land disposal facilities in the
United States, according to a 1965 APWA survey,[3]
was estimated at 1,000 to 1,250 non-captive sanitary
landfills, and 17,500 to 21,300 non-captive open
dumps. Non-captive installations are those not
operated for the disposal of the owner's refuse
exclusively. The survey covered 19,790 locations
(6,041 urban, 4,141 semirural, and 9,598 rural).
The ratio of open dumps to sanitary landfills was
calculated as an average of 94 to 6. The subtotals
show ratios of 5,760 to 934 open dumps to sanitary
landfills for urban areas; 4,110 to 41 for semirural;
and 9,502 to 96 for rural. Of the total number of
sanitary landfills, the urban areas supported 87.2%,
the semirural 3.8%, and the rural 9.0%.

Capital investment needs for the U.S. for landfill operations for the period of 1966-1975 are
projected at $340 million out of a total of $2.42
billion (1965 dollars) for all waste disposal
facilities including collection, storage, and maintenance. In estimating the sanitary landfill capital
investment needs, it is assumed that open dumps will
be eliminated wherever feasible and converted to
sanitary landfills. Such conversions are estimated

to require $200 million for the decade. As a matter
of interest, EPA has as one of its immediate goals,
"Mission 5000," the elimination of 5000 open dumps
in favor of pollution-free and aesthetically
acceptable methods. Sanitary landfill would be one
such method.

Open and burning dumps are a disgraceful and
unnecessary part of our nation's environmental crisis.
Such dumps contribute to air and water pollution;
provide food and harborage for rats, flies, and other
pests; are potential sources of disease and accidents;
and are just plain ugly. Unfortunately, open dumps
are also the most common means of solid waste disposal
in the United States.

A survey by various states in cooperation with
the Office of Solid Waste Management Programs, EPA,
indicated that only 6% of all authorized land disposal sites (those used by regular collection services,
meet accepted standards. Nearly half of such open
dumps contribute to water pollution and three-fourths
to local air pollution. In contrast, the sanitary
landfill avoids problems of environmental pollution
and frequently results in the creation of valuable
new land for parks and other recreational areas.

The success of "Mission 5000" depends upon dedi-
cated action by officials at federal, state, and
local levels; the encouragement of civic, trade,
and professional organizations; and the understanding
and support of every citizen.

The role of the federal government in "Mission
5000," though important, is limited. The OSWMP offers
to render technical assistance, including furnishing
recommended standards and model legislation. Special
training courses in solid waste management will be
offered for operators, supervisors, and public
officials; with the help of state and local personnel,
OSWMP will monitor the progress of "Mission 5000."

OSWMP considers that regulatory authority to
ensure proper collection and disposal of community
wastes is strictly a state or local responsibility.
Accordingly, full support and participation by state
and local officials are essential.

REFERENCES

1. Brunner, D. R., D. J. Keller, C. W. Reid, and J. Wheeler.
   "Sanitary Landfill Guidelines--1970," review draft of a
   Public Health Service publication, Office of Solid Waste
   Management Programs, Environmental Protection Agency.

2. American Public Works Association. *Municipal Refuse Disposal*, 3rd edition. Interstate Printers and Publishers, (1970).
3. American Public Works Association. "Solid Wastes--The Job Ahead," highlights of a report by APWA Research Foundation, *APWA Reporter*, pp. 5-11, 25 (August, 1966).
4. Warner, A. J., C. H. Parker, and B. Baum. "Plastic Solid Waste Disposal by Incineration or Landfill," study prepared for Manufacturing Chemists Association (December, 1971).

PART IV

PLASTICS IN INCINERATION
AND LANDFILL PROCESSES

CHAPTER 21

PLASTIC WASTE COMPOSITIONS--
PAST, PRESENT AND FUTURE

UNITED STATES

While considerable discussion is taking place,
in the United States particularly, as to the means
for adequately handling the solid waste loads that
an affluent and growing urban and suburban society
is generating in increasing amounts, the fact remains
that landfill is still the most widely practiced
method, with incineration increasing but still in
rather small proportion (about 10%), of the total
load. Sanitary landfill (as opposed to dumping),
while itself accounting for only a portion of the
total amount of solid wastes disposed by landfill,
is a very satisfactory method of disposal.

Incineration, properly performed, provides the
most advantageous approach to the basic problem
posed in solid wastes management, namely, how to
rapidly and effectively reduce the large volume
generated to the smallest, sterile residue that can
be disposed with minimum further effect on the environment. Present advances in incinerator design
and operation, coupled with other developments in
solid wastes handling, are making the technical
answers more readily available. But these advances
are not apparently matched at present in the U.S.A.
by the necessary socio-political willingness to
assume general acceptability of the thesis that,
particularly for the larger municipalities, large
scale incineration with heat recovery and adequate
air pollution controls is, in fact, the most satisfactory route available now and for the foreseeable
future.

In this chapter, the discussion concerns what is
the present and expected future role of plastics in
the solid waste stream and what, if anything, needs

to be done to insure that the quality of life is not
adversely affected by the continued and expanded use
of such materials. To do this it will be necessary
to consider the question of what constitutes the
solid wastes that have to be disposed, how the composition of these wastes has changed with time, and
the varying nature of the present and expected loads.
With this knowledge it will then be possible to
examine what happens when such wastes are either
landfilled or incinerated, examine the pressures of
a concerned society for an improved environment,
begin to understand more fully the arguments of the
proponents of this or that particular approach to
the problem, and comprehend the difficulties of
determining any one specific answer that can command
universal acceptance.

Thirty-five years ago, the average solid wastes
were predominantly garbage mixed with a certain
amount of paper and paper products, cans and bottles,
mixed with ashes, dust and cinders. They had a relatively high density and low calorific value. In
contrast, today's solid wastes, especially those from
urban and suburban areas, are high in paper and paper
products and relatively low in garbage and ashes,
with a corresponding and constantly decreasing density coupled with an increasing calorific value.
While plastics materials are still a relatively small
percentage of the solid wastes load (in fact, in many
statistics dealing with this subject plastics are
still not separated out as a category but often shown
generally within "plastics, rubber, leather, etc."),
they contribute in quite a substantial way to the
questions posed by this increase in calorific value
and decreased density. To many investigators already
concerned with a desire to utilize the heat values
in refuse, the increasing calorific value and the
apparent stabilizing of the value at a higher level
than heretofore present an ideal reason for the
adoption of incinerators using waste heat recovery
systems. To them, the plastics content in the refuse,
both at the present and the expected higher levels of
the future, are an attraction in many ways. They do,
however, need to be reassured that the nature of the
combustion products coming from the increased plastics
materials content of the refuse will not basically
change from those seen today.

To those owners and operators of incinerators
not having the most up-to-date equipment, the increased calorific value of the present-day refuse is
already causing some concern, especially in the area

of air pollution control, and an increasing plastics materials content adds to their concern. On the other hand, it is generally admitted that other concerns have lessened, principally the ability to keep a furnace fire burning, even during the wettest of months, and the maintenance of a more uniform burn-out of material. In this case, plastics materials have been considered as particularly advantageous, with some operators actually separating out plastics materials from the incoming refuse for use during periods of especial difficulty. In this respect, it might be briefly mentioned that the use of plastic garbage bags is seen by most municipal authorities and landfill and incinerator operators as a very positive aid to the disposal question because the bags facilitate the handling of trash.

For the present day designer of incinerators and the municipal planner of landfill sites, whose equipment and facilities are expected to perform satisfactorily over at least a 25-year period, a more detailed understanding of the present and expected composition and characteristics of the solid wastes is therefore quite essential. While a considerable amount of data has been published on this point, obviously reflecting the importance of the entire subject, it is somewhat unfortunate that there still remains a lack of complete agreement among the various experts as to the classification of the actual solid wastes, thereby making it quite difficult to correlate the data from the various sources. An attempt at correlation was made by Niessen and Chansky in their report to the 1970 National Incinerator Conference.[1] They searched all available information sources, including demonstration, research and planning grants, and data directly generated by the Office of Solid Waste Management Programs, and uncovered only 23 useful, apparently independent data sets. These data may be compared with other indicative data compiled by the American Public Works Association in 1966.[2]

Niessen and Chansky comment on the wide variation in average composition reported and find two particular categories--"yard waste" and "miscellaneous"--particularly highly dependent on local conditions. The fraction of "yard waste" seems very sensitive to both geographical location and the season of the year (a not altogether unexpected finding) while "miscellaneous" is dependent on local practices and regulations that are concerned with the collection of demolition and other such wastes. The data were

therefore modified to reflect an average composition
of municipal wastes with the yard wastes and miscellaneous materials removed. This is given in Table
21.1.

Table 21.1

An Average Waste Consumption

Component	Mean Weight, %
Glass	9.9
Metal	10.2
Paper	51.6
Plastics	1.4
Leather, Rubber	1.9
Textiles	2.7
Wood	3.0
Food Wastes	19.3
TOTAL	100.0

Using data scanned,[1,2] a 30-year comparison can be
made as follows:

Component	1939	1953	1968
Glass	5.4	8.5	9.9
Metal	6.7	11.0	10.2
Paper	21.9	48.0	51.6
Plastics	0	0.4	1.4
Leather, Rubber	3.2	4.8	1.9
Textiles			2.7
Wood	2.6	2.6	3.0
Food Wastes	17.2	24.7	19.3
Ashes	43.0	0	0
TOTAL	100.0	100.0	100.0

Recognizing the difficulties in attempting to
compare data of this type with any degree of certainty, it is still clear that a change in composition
of refuse has occurred from 1939 to the present time.
This can be seen particularly in the major disappear-
ance of ashes and their replacement by combustible
material such as paper and plastics, and increases
in other inorganic materials such as glass and metals.

Since 1953, the effects of prepackaged foods and other commodities is seen. The garbage content has dropped, more in urban than rural areas, paper and paper products have increased significantly, and glass and plastics usage has gone up substantially. These changes have been accompanied by a decrease in density of the solid wastes, and an increase in calorific value. Warner, Parker and Baum[3] have given data for the density of solid wastes containing the higher amounts of ashes seen some 30 years ago in New York City, for example, at 37.5 pounds per cubic foot. In 1966 the average for the United States was estimated at 13 pounds per cubic foot, with a maximum of 26 pounds and a minimum of 11.1 pounds. It is probable that today the average is closer to 11 pounds per cubic foot.

These data compared well with the data for the United Kingdom (12-14 pounds per cubic foot) and the Seine region of France (12.3-15.6 pounds per cubic foot). With the projected changes in solid wastes composition for the future (assuming no dramatic deviations such as a banning of packaging), densities as low as 10 pounds per cubic foot may be expected. The physical problems of transportation of the solid wastes could become most burdensome, and the effects on sanitary landfill (in the absence of any precomminution) quite marked.

It is apparent that with the rather large variations in composition that can occur from area to area and from season to season, any attempt to closely define the heat values of the solid wastes to be incinerated would be useless. However, this parameter is of vital importance to the incinerator designer.

Bowerman in Corey's book "Principles and Practices of Incineration"[4] gave a useful table, taken from a study by Purdue University of the components in an average municipal refuse as shown in Table 21.2.

The average quoted calorific value of 6203 BTU per pound is too high, as seen in data by DeMarco *et al.*[5] who state that present incinerator designers are using gross heat values ranging from 3,000 to 6,000 BTU per pound based on waste as received. The present trend indicates that heat values of incinerator solid waste will increase by 500 BTU per pound by 1980. A more recent study by Kaiser and Carotti[6] gives a value of 4,577 BTU per pound for a typical municipal refuse, with an additional 150 BTU released by partial oxidation of the metals planned, or a total of 4,727 BTU per pound.

Table 21.2

Municipal Refuse Components[4]

Component	Percent of All Refuse by Weight	Percent of Moisture by Weight	Calorific Value BTU/lb
Paper	42.0	10.2	7,572
Wood	2.4	20.0	8,613
Grass	4.0	65.0	7,693
Brush	1.5	40.0	7,900
Greens	1.5	62.0	7,077
Leaves	5.0	50.0	7,096
Leather	0.3	10.0	8,850
Rubber	0.6	8.2	11,330
Plastics	0.7	2.0	14,368
Oils, Paints	0.8	0.0	13,400
Linoleum	0.1	2.1	8,310
Rags	0.6	10.0	7,652
Street Sweepings	3.0	20.0	6,000
Dirt	1.0	3.2	3,790
Unclassified	0.5	4.0	3,000
Garbage	10.0	72.0	8,484
Fats	2.0	0.0	16,700
Metals	8.0	3.0	124
Glass & Ceramics	6.0	2.0	65
Ashes	10.0	10.0	4,172
All Refuse	100.0	20.7	6,203

As indicative of the values in earlier years when ashes were present to an appreciable extent, the 2,700 BTU per pound reported for Hilsheimer[7] in the Mannheim, West Germany incinerator for a refuse containing 27% of ash and noncombustibles may be cited here.

Looking now at the plastics materials which are entering the United States' solid wastes stream now, or expected in the future, the calorific values will depend somewhat on the exact formulation, but those now present fall within the following ranges.

Material	*BTU/lb*
Average solid waste	3,000- 6,000
Polyolefins (PE, PP, etc.)	18,500-19,500
Polystyrene	17,000-17,500
Polyamides (nylon)	12,000-13,500
Polyesters	11,500-12,500
Polyurethane	11,000-12,000
Polyvinyl chloride	7,500- 9,000

In actuality, the calorific value can be calculated from the composition of the plastics material itself, and so a reasonably accurate picture of the future in this regard can be derived.

It should be recognized that a real accomplishment in the drive to reclaim and reuse metals and glass, which normally enter the solid wastes stream for ultimate disposal by incineration, could be a new phenomenon causing a modification of the observations above. In a system where the solid wastes are first comminuted and then magnetically, electrically and photometrically sorted to remove the majority of the metals and glass, with the remainder going to the incinerator, there would be in the refuse input the removal of a substantial proportion of the low heat value material, and consequently a higher heat value than those presently observed. Also a major increase in handling household, commercial and industrial refuse in plastic bags could lead to a substantially lower moisture content as received at the incinerator site and thereby an increase in calorific value. All of these effects are readily calculable, and the general approaches of Kaiser based on the well known Du Long formula are applicable. Thus: gross BTU per pound = 145.4 C + 620 (H - O/8) + 41 S, where C, H, O and S are the weight percentages in the refuse of carbon, hydrogen, oxygen and sulfur, respectively, whether as solid or liquid.

WESTERN EUROPE AND SCANDINAVIAN COUNTRIES

Attention to the role of plastics in an overall evaluation of environmental questions generally has proceeded steadily in Europe for some years now, with particular emphasis on the technical questions posed by an ever increasing change, both in amount and in composition of municipal wastes. These changes have come about because of two major trends

over the past decade. The first relates to changes
in methods used for heating homes and office buildings
where traditional coal heating has been gradually
replaced by oil and gas heating. This has lead to
a marked diminution in the amount of substantially
inorganic ash in the municipal wastes. The second
change relates to the increasing use of all types
of convenience packaging which has markedly increased
the amounts of organic matter (cellulose in paper,
plastics, wood and so forth). Both trends have
resulted in an increase in the calorific value and
a decrease in the density of municipal wastes. The
increase in calorific value has led to the necessity
for closer control and understanding of incinerator
design and operation, while the decrease in density
complicates the collection problem as well as disposal either by sanitary landfill, incineration, or
other disposal methods.[3]

Rasch[8] gives the following data for the composition of domestic waste in West Germany over the past
few years, as well as projections for the future.

	1965	1970	1975*	1980*	1990
Ashes, dust, cinders	60	32	25	18	5
Paper and cardboard	14	40	45	50	60
Organic garbage	18	17	16	15	13
Metals	3	3	3	3	3
Glass	3	4	5	6.5	10
Plastics	1	3	4	5.5	8
Textiles	1	1	1	1	1

*Interpolated from 1970 and 1990 projections.

For comparison, Staudinger[9] shows the following
data for the United Kingdom:

	1955	1960	1965	1975
Ashes, dust, cinders	53	51	36	18
Paper and cardboard	15	16	23	50
Organic garbage	12	12	17	13
Metal	6	6	7	6
Glass	6	6	8	6
Debris (including plastics)	6	6	6	4

A further factor noted by all countries is the increasing *per capita* generation of solid waste for ultimate municipal disposal, and the relative uniformity of both amount and composition of such wastes among the more technically advanced countries. While there are still differences between countries as to the per capita generation of solid wastes, all agree that the total amounts are increasing at a marked rate, accompanied by a rapid increase in volume as discussed earlier. Thus Rogus,[10] in discussing the question of refuse incineration in Western Europe, derived some figures from the data of seven large cities showing that the refuse output rose from 215 kgs per capita per year in 1953 to 318 kgs per capita per year in 1963. He estimated that in 1966 the figure would be 330 kgs per capita per year.

Sweden reports by one authority[11] the quantity of household refuse in the 1940's as 190 kgs per person per year, 300 kgs in 1962, and estimates 470 kgs in 1970; by another authority, 1968, 400 kgs; 1970, 430 kgs; 1975, 500 kgs; and in 1980, 600 kgs per person per year. Germany reports 240 kgs per person per year for 1968, plus or minus 25% for any individual month.

France saw an increase in total household waste of 15% between 1961 and 1965. In a relatively recent report[12] Rousseau reports on the Department of the Seine, which includes Paris and more than 80 suburban municipalities, for a total of some 4,800,000 inhabitants. The average per capita per year value is given as 333 kgs.

The excellent and recent work of the Society of the Chemical Industry in their Monograph Number 55, *Plastics Waste and Litter*,[9] reports on the present situation and projections for the future. Their data for the United Kingdom is given as 294 kgs per capita per year for 1968, with estimates of 362 kgs in 1975 and 408 kgs in 1980. There will be occasion to deal more extensively with the subject matter of this monograph later in this discussion.

The increasing per capita generation of solid waste requiring attention at the municipal level has been marked by a rapid increase in all types of packaging materials. While plastics has been a very visible contributor in this area, it should be noted that paper, cardboard and wood have substantially increased in volume also. Thus, in Europe, there is little disagreement that the actual percentage of plastics in the municipal wastes at the present time

is relatively low--2 to 4%--but that, with the
present upward trend in the housewife's preference
for convenience packaging, this figure may be expected
to continue to increase in the next decade.

For the United Kingdom, Saudinger[9] gives the
following statistics for plastics:

1. The major proportion of plastics waste originates
   at present from the packaging sector.
2. Practically all plastics packaging is becoming waste
   within a comparatively short period of use-life, and
   nearly all will appear as part of the collectable
   domestic and trade refuse.
3. A certain amount of plastics other than the packaging
   items will become dust bin waste. This amount is at
   present estimated to be 10% of the nonpackaging
   applications.
4. The amount of plastics waste from applications outside the packaging area and its proportion in relation
   to the total plastics waste will increase from year
   to year.

With the year 1968 as a base, the statistics for
the United Kingdom are given as:[9]

Population (including N. Ireland)	55,390,000
Production of plastics	1,220,000 tons
Consumption of plastics	1,105,000 tons
Plastics (use in packaging)	250,000 tons
Packaging as % consumption	22 %
Additional plastics waste (calculated)	85,500 tons
Per capita per year	44.5 lb
Per capita per year	9.8 lb
Refuse collected	16,000,000 tons
Per capita per year	647 lb
Plastics in collected refuse (found)	1.12 %
Plastics in collected refuse (found)	180,000 tons
Per capita per year	7.3 lb
Plastics waste (calculated)	335,500 tons
Plastics waste (calculated)	2.1 %
Per capita per year	13.5 lb
Plastics packaging in refuse	1.56 %

The 250,000 tons of plastics used in packaging
compares to a total 1,105,000 tons of plastics used
for all purposes, including packaging. Thus by
using Staudinger's assumption of 10% of all nonpackaging applications of plastics reaching the
disposal point each year, a total calculated plastics
waste of 335,500 tons, or 2.1% of the total wastes

of the country is reached. In finding, by actual count, 1.12% plastics waste in the total garbage collected, it should be remembered that the actual count was based on averages taken over several collections from the city of Birmingham (a Midlands city of some one million population) where it was also found, as previously noted for other European experiences, that quite substantial differences exist from day to day.

As would be expected, only a comparatively small number of types of the total of plastics materials available make up the considerable bulk of plastics packaging materials found in waste. These are the major thermoplastics, the polyolefins, the polystyrenes, and the PVC compounds. A breakdown of the percentages used for packaging applications is as follows:

Polyolefins (including 8% polypropylene)	74%
Polystyrenes	16%
PVC	4%
Thermosetting materials	3%
Miscellaneous	3%

The plastics wastes appearing in the disposal system consist primarily of films, wrapping foils, bags, sacks, bottles, tubs, cups and beakers where the surface area to weight is very large. A smaller percentage exists for thicker walled products such as cans, canisters, small drums, kegs, box inserts, large bottles, trays, punnets (shallow boxes used for shipping fruits or flowers) and closures.[9]

Future Plastic Waste

In discussing the future quantities of plastics for disposal, the SCI report[9] arrived at the following table for the United Kingdom based on growth factors of 10% and 12%. The data are in Table 21.3.

It is observed that even at the highest assumed rate of increase per year (12%), there would still be a 5- or 6-year time lag before per capita consumption in the U.K. equaled 1969 levels in the U.S.A., Sweden or Germany. Thus their forecast, at least for 1975, would seem to be supportable by analogy between the already established product and application patterns of these latter countries.

Table 21.3

Plastics Future Consumption

	1970		1975		1980	
	10%	12%	10%	12%	10%	12%
Production, in 1000 tons	14.75	15.30	22.40	27.00	36.10	47.50
Consumption, in 1000 tons	12.54	13.00	19.05	23.00	30.48	41.58
Population, million	55.97	55.97	57.58	57.58	59.25	59.25
Per capita consumption, lb	50.0	51.2	72.8	87.8	115.0	154.0

Looking into the future for the individual plastics materials making up the total for 1975 and 1980 at an assumed growth rate of 10%, the SCI report[9] expects similar rate of growth for the next three years with a tailing off towards the second half of the decade. An overall growth rate of around 12% does not appear overly optimistic and can reasonably be expected.

It is perhaps of interest to compare the plastics packaging ratios in countries where the per capita consumption of plastics in 1968 was higher than in the U.K. and to which packaging must have contributed:

	U.K.	Japan	U.S.A.	Germany
Per capita consumption, lb	44.5	48.0	72	92
% used for packaging	22	20	18	17
Per capita use of packaging, lb	9.8	10.5	25	15.5

It is immaterial here whether or not this relationship is significant, but what these figures indicate is that other application areas such as building, transport, motor cars, household appliances and gadgetry for sports, leisure, and education have in the past recorded a higher growth rate than packaging.

They also show that as far as packaging is concerned
plastics materials in the U.K. still have some way
to go to reach standards comparable with other
countries, and thus again an average growth rate of
12% does not seem overly optimistic. Based on this
rate of growth the packaging industry would require
a minimum of 900,000 tons by 1980, and as in the
earlier years of the decade there will probably be
a higher growth rate. It is estimated that in 1975
the total required will be of the order of 500,000
to 600,000 tons. How will this affect the plastics
content in the refuse? To find the answer to this,
one will have to know the total plastics waste
reaching the disposal site and the total amount of
refuse collected for disposal. The plastics waste
originates from three sources: packaging uses
(dustbin waste); nonpackaging applications (of
dustbin "quality"); packaging and nonpackaging
articles (larger than dustbin quality). As far as
packaging waste is concerned, one can assume that
at least 85 to 90% becomes collectable waste, with
the rest being retained, reused, or disposed of by
"do-it-yourself" methods or by litter-dropping.
However, this quantity of packaging waste, although
it is the major contribution to the plastics content, is not the sole source of waste in the refuse.
To this must be added all the other discards, rejects and unwanted household items which find their
way into the dustbin.

Extending per capita refuse production to the
whole of the U.K. population, the total refuse in
1968 would amount to 15.8 million tons. This figure
reflects only material known or estimated to be
collected.

When predicting the future production of refuse
and the quantities which have to be disposed of, some
uncertainties arise not only as to the rate of growth,
but also in respect to compositional changes because
of the extraction of salvageable materials.[9] The
aforementioned source quotes two predictions, one
relating to Coventry where future refuse is expected
to increase by 50% by weight over the next 20 years
and the other referring to London's refuse where an
increase of 67% over the next 20 years is anticipated.
In both cases the volume is expected to double during
this period. Assuming a straight-line progression
of the increase between 1963 and 1983, refuse production for the whole of the U.K. would be as follows:

	1968	1975	1980 (at 50% growth)	1980 (at 67% growth)
U.K. total, million tons	15.8	20.5	22.5	25.0
Per capita/year, lb	646	797	850	945
Per house/week, lb	37.5	46.0*	48	55

*37 lb has been quoted in *Public Cleaning* by F. Flintoff and R. Millard.

Results from individual locations like Birmingham or London, where the refuse is also a mixture of domestic or trade origin, show that the urban refuse is at present somewhat lower than the above average, *i.e.*, 29 to 37 pounds per dwelling per week.

In line with the increase of paper products in refuse, mainly due to packaging, the plastics content is also increasing; information about the level it will reach by 1980 is of considerable importance, particularly for the planning of future incineration facilities. Owing to its high heat value, the plastics content is a parameter which is vital for the correct design and successful operation of incinerators, and hence any information on this point should be as realistic as possible. Although one can compute the total plastics waste with a fair degree of accuracy, one cannot assume that all the waste will appear within the collected refuse; it is this point which introduces a degree of uncertainty.

To put plastics packaging waste figures into perspective, one might consider them in relation to other "soft" packaging waste, *i.e.*, paper-based packaging materials. These products are expected to grow at an annual rate of about 5%, the rate at which they have grown over the last five years, so that by 1980 their potential waste quantity would amount to about 5.5 million tons. Even without taking into account other packaging wastes, *i.e.*, glass, metal, etc., the share of plastics in packaging and in waste generation remains quantitatively a minor one.

Composition Data

The composition of the packaging waste will be in line with the product distribution within the packaging applications, whereas the nonpackaging waste does not follow any application pattern but in time will approach a composition corresponding to the ratios of all the plastics which are available for nonpackaging uses. These two waste-producing areas have been calculated on the above two assumptions (Table 21.4, lower limit in left-hand column).

Table 21.4

Composition of U.K. Plastics Waste in the Future[9]

	Low (tons)	High (tons)
Packaging waste		
Polyolefins	500,000–550,000	575,000–600,000
Polystyrenes	150,000–175,000	130,000–180,000
PVC	80,000–100,000	108,000–120,000
Cellulosics	8,000– 10,000	11,000– 12,000
Miscellaneous	22,000– 25,000	26,000– 38,000
Total	760,000–860,000	850,000–950,000
Nonpackaging waste		
Polyolefins	92,500	130,000
Polystyrenes	35,000	50,000
PVC	102,500	140,000
Miscellaneous	29,000	40,000
Thermosetting compounds	101,000	140,000
Total	360,000	500,000
Composition of total plastics waste		
Polyolefins	592,000–642,500	705,000–730,000
Polystyrenes	185,000–210,000	180,000–230,000
PVC	182,500–202,500	248,000–260,000
Miscellaneous	51,000– 54,000	66,000– 78,000
Cellulosics	8,000– 10,000	11,000– 12,000
Thermosetting compounds	101,000–102,000	140,000–140,000
Total	1,120,000–1,220,000	1,350,000–1,450,000

Some specific problems connected with the European waste situation[3] were brought out by various authors at an International Symposium on Disposal of Plastics Waste and Litter which was held in Oslo, Norway in May, 1970. The opening address by Mr. Helge Seip, Norwegian Minister of Municipal and Labour Affairs, indicated among other things that incineration was the most feasible method of solid waste disposal for Norway. Dr. Schonborn of West Germany (BASF) reported

that their incinerator successfully handled wastes
containing more than 10% PVC. It is estimated that
by 1980 the plastics content of refuse would be 6%,
of which 15 to 20% would be PVC. Therefore, successful operation of an incinerator at a PVC level six
times that expected by 1980 has already been
accomplished.

Dr. Wogrolly of Austria discussed the decomposition of plastics and the resulting gaseous products.
From his detailed information, one important conclusion was a reemphasis that no phosgene was detected
in the gaseous decomposition products of PVC.

The Royal Institute of Technology, Stockholm,
set up a study to "uncover the facts of plastic
waste disposal" after wild and inflammatory accusa-
tions by the press, as well as student groups,
concerning the "evils" of PVC. The press filed
stories about the poisonous phosgene fumes and the
poisonous and corrosive HCl fumes from PVC incineration. The paper presented by Professor Bengt Ranby,
a result of this study, indicated the miniscule role
that HCl evolution from PVC plays in air pollution.
Figure 21.1 taken from this paper shows dramatically
the relationship between acid precursors hydrogen
chloride and sulfur dioxide.

A paper by Hans Aeberli suggested that biological
degradation could not be considered as a means of
disposal for plastics except, perhaps, for cellulosics.
He indicated that by 1980 the plastic content of refuse
in Germany would double to 6% with PVC remaining at
1/3 of this. He discussed the various disposal
methods including detailed analyses of domestic refuse.
His conclusions on incineration are as follows:

"It is, therefore, not expected that in the foreseeable
future the amount of plastics in domestic refuse will reach
such proportions as to cause difficulties in normal incineration
plants with grates. A move towards other incineration facilities such as the revolving drum is totally uncalled for; on
the contrary, full-scale tests carried out with refuse to which
plastics were added in ever increasing proportions up to 12%
proved clearly that the revolving drum is not at all suited to
this purpose. The carbon content in the residue was found to
increase with the proportions of plastics. The reasons for
this are manifold, to mention only one, the previously mentioned flowing properties of the clinker, due to which the
lumps of clinker are constantly covered with liquid plastic
and as a result, the oxygen necessary for successful burn-out
of the fixed carbon cannot gain access. The best principle
would be to continue separating domestic from industrial refuse

in cases where the proportion of plastics is abnormally high, and to dispose of large amounts of single-kind and similar plastic waste in special industrial refuse incineration plants, where the revolving drum would be justified."

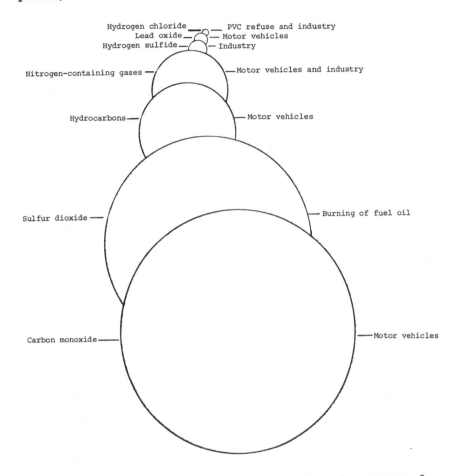

Figure 21.1. *Comparison between amounts of various types of atmospheric pollution in Sweden.*[13]

JAPAN

The Japanese technology in the matter of solid waste disposal is similar, as might be expected, in most respects to that of the United States and Western Europe. The major differences in Japanese technology from that of the United States are:[3]

The rate of plastics production in Japan is expected to increase more than fivefold, 1966 to 1975. The percentage of plastics in total solid wastes in 1968 was nearly 7%, or more than three times that of the United States. Further, it is estimated that of the total plastics produced in Japan in 1970, about half will find its way into the channels of solid waste for disposal.

The Tezuka Kosan Refuse Compression Process appears to be the only pilot compression process developed to approximately production scale. In this process, raw refuse is placed in a ram press generating intense pressure which compresses and partially dehydrates the mass. The compacted mass is then placed in a chicken wire or thin sheet steel and encapsulated in asphalt or concrete. The end product is a solid block which can be used in construction, building dikes for flood prevention, and in other ways.

Refuse in Japan has a far higher moisture content, a lower calorific value, a higher percentage of garbage, a higher percentage of plastics, and a lower quantity of paper than that of the United States or Europe. The basic Japanese problem is the development of incineration technology suitable to their high moisture, low calorific value refuse, coupled with relatively high PVC content.

In industrial waste disposal, including plastics, the Japanese government considers that industry is responsible for its own wastes. Most of the waste is taken care of by contracting with outside disposal services at about 5 cents a pound. Like other countries, Japan is running out of suitable space for sanitary landfill. Disposal of solid waste, however, is mostly by landfill, with increasing amounts by incineration.

The Japanese government is sponsoring some effort in the burning of plastics for fuel and is organizing a program of cooperation with industry in the control of air and water pollution from solid waste operations. Since approximately 35% of all Japanese plastics are vinyl chloride resins, the Japanese are also highly sensitive to corrosion problems.

What has been obtained in the matter of figures for Japan is shown in Table 21.5. The future statistics for Japan show that the following points are pertinent:[3] The total production of plastics in Japan has increased amazingly since 1955, and it is estimated that 4 million metric tons were turned out in 1969. Further, plastic production has increased

Table 21.5

Plastics Waste in Japan

Year	% of Plastics in Total Waste Volume	Plastics Produced (In Thousands of Metric Tons)	Solid Waste Plastics
1958	-	245	90
1961	-	690	260
1963	2.2	-	-
1964	4.1	1,377	588
1965	4.3	-	-
1966	4.8	1,994	894
1968	6.7	3,400	1,500
1970 (est.)	-	5,300	2,300
1973 (est.)	-	7,850	3,750
1975 (est.)	-	10,000	5,100

at an average of 30% yearly during the past ten years, and it is believed that Japan is now the world's second largest plastics producer. All sorts of containers, dishes, packaging materials, bags, toys and functional parts are all markets for PVC, polyethylene, polystyrene and expanded PS, polypropylene and others because of their resistance to corrosion and cracking as compared with metal and wood.

According to early 1970 statistics, only 3.3% of plastic materials were in Japanese solid waste in 1963. This jumped, in Tokyo, from 6.5% in 1965 to nearly 12% in 1968! In Yokohama, plastic materials in solid waste increased from 4.5% in 1965 to 8% in 1968. Similarly, in Kawasaki, the increase was from 7% in 1965 to 9.1% in 1968. These figures show that the amount of plastic refuse has almost doubled within the past five years and has become about 10% of the total municipal solid wastes.

The situation in Japan has many parallels with that in the United States. For example, there is a keen awareness of the problems and the necessity for combining the efforts by the governmental bodies and private industries for working out the problems of plastics solid waste disposal.

On the assumption that there has been presented
a reasonable picture of the composition, heat content
and density of the refuse, and that one can reasonably
estimate the changes to these figures based on any of
a number of possibilities in the future (increase in
plastic bottles and decrease in glass bottles--prior
removal of metal and glass before landfill or incineration, for example), there will next be an examination of the specific role of plastics materials in
the two methods of disposal currently under review.

REFERENCES

1. Niessen, W. R. and S. H. Chansky. "The Nature of Refuse," paper prepared for 1970 National Incinerator Conference (ASME) at Cincinnati (May, 1970).
2. American Public Works Association. *Municipal Refuse Disposal*, 2nd edition (1966) and 3rd edition (1970), Interstate Printers and Publishers.
3. Warner, A. J., C. H. Parker, and B. Baum. "Solid Waste Management of Plastics," a study prepared for Manufacturing Chemists Association (December, 1970).
4. Corey, R. C. *Principles and Practices of Incineration,* J. Wiley and Sons, Inc. (1969).
5. DeMarco, J., D. J. Keller, J. L. Newton, and J. L. Leckman. "Incinerator Guidelines, 1969," Public Health Service Publication Number 2012, Office of Solid Waste Management Programs, Environmental Protection Agency.
6. Kaiser, E. R. and A. A. Carotti. "Municipal Incineration of Refuse with 2% and 4% additions of Four Plastics: Polyethylene, Polyurethane, Polystyrene and Polyvinyl Chloride," a report to the Society of the Plastics Industry (June, 1971). Also in a paper prepared for 1972 National Incinerator Conference (ASME) at New York (June, 1972).
7. Hilsheimer, H. "Experience after 20,000 Hours--The Mannhein Incinerator," paper prepared for 1970 National Incinerator Conference (ASME) at Cincinnati (May, 1970).
8. Rasch, R. *Energie*, Vol. 21, pp. 7-8 (1969).
9. Staudinger, J. J. P. "Plastics Waste and Litter," SCI Monograph Number 35, published in the United Kingdom by Staples Printers Ltd.
10. Rogus, C. A. "An Appraisal of Incineration in Western Europe," paper prepared for 1966 National Incinerator Conference (ASME) at New York (May, 1966).
11. Report by Royal Swedish Academy of Engineering Sciences, Communication Number 160, Stockholm (1969).
12. Rousseau, H. "Large Plants for Incineration of Refuse in the Paris Metropolitan Area," paper prepared for 1968

National Incinerator Conference (ASME) at New York (May, 1968).

13. Ranby, B. "Plastics from an Environmental Protection Point of View," Royal Technical University, Institute of Polymer Technology, Stockholm, Sweden, EFTA Conference, Oslo, Norway (May, 1970).

CHAPTER 22

PLASTICS IN INCINERATION

As has been stated before, municipal authorities and operators of incinerators now have a reasonably accurate picture of the range of density, composition, and the spread of heat values of the product to be handled. The plastics industry can, with a fair degree of certainty, assure them that the future will not bring with it any basically new problems as far as their contribution to incineration is concerned, and that adequate technology exists for the proper disposal of the quantities of plastics that now enter the solid wastes stream or are expected in the next decade.

Regarding the present performance of plastics in the solid waste stream, a general cross-section of comment would include the following:

a. "We find the presence of plastics in the refuse very helpful in keeping the fires burning, especially when it has been raining heavily and the refuse is set."

b. "I know some people are saying that plastics give troubles in incinerator operation, but quite frankly, for my installation, I have never seen this."

c. "The only time we have trouble with plastics is when a truckload of refuse arrives from an industrial operation which is mainly plastic, gets dumped in the pit and the crane operator picks up a load without mixing. When this hits the fire, we get a flare-up, it is difficult to control the burning, and a large smoke evolution occurs till the situation corrects itself."

d. "From my own observations, I can honestly say that the difficulties of incinerator operation and maintenance are no different now from several years ago, when plastics were not even mentioned as being in the refuse."

Obviously there are some municipal engineers and
operators of incinerators who ascribe certain difficulties to the increasing amounts of plastics materials
seen in the refuse, and these alleged difficulties may
generally be classified as:

    a. "Plastics in the refuse causes sudden flare-ups on
the grates which give an increase in particulate
matter emission and contribute to premature burn-out
of the grates."

    b. "Plastics in the refuse melt on the grates during
burning, drip through causing undergrate fires and
smoky operation, and interference with the proper
flow of air through the burning bed."

    c. "Should there be any appreciable quantity of plastics
in the refuse being incinerated, the plastic melts,
coats the other refuse, and tends to cause the burning
bed to slide down the inclined grate. This leads to
too short a residence time in the furnace and an
inadequate burn out."

It would be unwise to ignore or summarily dismiss
these charges, exaggerated though they might be in
many cases, Especially considering some of the obsolete incinerators in use today and the lack of attention by operators in certain instances, these problems
may occur. When analyzed more closely. Items (a),
(b) and (c) may all be ascribed to a lack of homogeneity
of the refuse entering the incinerator furnace. This
question has been of concern to designers and operators
of incinerators for many years and has become increasingly important with the decreasing density of
the refuse. Should operating practice permit the
discharge into the collection pit of trucks containing
a large proportion of plastics scrap with no special
effort by the crane operator to feed the furnace by
judicious removal of refuse from the various areas of
the collection pit, thereby effecting a mixing, it is
obvious that a refuse containing appreciably more than
the average 2-3% plastics content can reach the
burning grate. For most types of moving grate incinerators, a plastics content greater than 20% could
give rise to this type of problem.

Adoption of the modern engineering concepts that
all refuse should be comminuted and mixed well before
incineration would certainly overcome many of these
difficulties. It may be noted that for several
reasons, and not just for the burning characteristics
alone, great attention is now being focused on this
aspect of the solid waste disposal problem. However,

it should be recognized that the comminution must
not be too great or the refuse will compact and not
burn steadily on the grates since the under-fire air
will be prevented from free passage through the
burning bed. It is believed that no serious problems
will exist up to at least 10% plastics content of the
refuse with conventional types of incinerators. For
wastes having greater than 10% plastics content, an
unlikely event for municipal incinerators ever,
special designs of incinerator systems are already
available, as discussed earlier on incineration.

The one item that therefore remains for serious
consideration is the charge that plastics on incineration give rise to active gases which cause corrosion
of the equipment, leading to frequent shutdowns, high
maintenance costs and short life. The alleged culprit
for this corrosion is, of course, polyvinyl chloride
(PVC) and its compounds which generate hydrogen
chloride on incineration. All other major plastics
materials are either hydrocarbon in nature or are
based on polymers generally containing carbon,
hydrogen and oxygen which, on burning properly, give
rise to carbon dioxide and water, both essentially
inert. There is a proportion of plastics materials
containing nitrogen--this special aspect is dealt
with later.

It is necessary, therefore, to pay especial
attention to this particular component (PVC) of the
refuse, despite its relatively low proportion, not
because of any really positive evidence to support
the claims of some that it is the villain causing
problems of corrosion in incinerators, emission of
unwanted gases into the environment and so forth,
but because its presence together with increasing
quantities of other potentially corrosive materials
does require that both the designers and operators
of the newer incinerators take the necessary, well
known steps to remove or minimize the potential
difficulties.

In a number of European countries where the
question of the effect of increasing amounts of
polyvinyl chloride entering the solid waste stream
for disposal by incineration has been well discussed,
this subject is treated in a more factual and non-
emotional manner. In Europe, there is even more
acceptance now than there was previously that the
presence of polyvinyl chloride in the refuse going
to incineration, even in the amounts forecast for
1980, poses no unacceptable restrictions.[1]

In the year 1971, the amount of polyvinyl chloride compounds in the U.S. solid waste stream going to incineration was probably averaging about 0.2 to 0.25%. This quantity consisted in large measure of plastic bottles, tubes, bags and other containers, some film and a smaller proportion of other articles such as floor tile, garden hose, plastic raincoats, scrap wire and cable. It is difficult to arrive at a definitive compositional analysis of these wastes, since they obviously contain other ingredients than the base resin. The effect is that the actual chlorine content is lower than at first expected. Thus, while the pure resin has a chlorine content of 56.8%, a typical bottle might only contain 49% chlorine, and a flexible raincoat or a wire and cable compound might be as low as 34%. This is well illustrated in the Kaiser and Carotti report[2] with an analysis of the actual polyvinyl chloride plastics used in the experiments performed. They state: "Five types of PVC bottles, from 6 to 24 fluid ounce capacity, PVC film, sheet and heavy molding scrap were received from processors."

Representative samples of the plastics mix were chopped in a Wiley mill to pass 2-mm round openings and were analyzed to give:

Moisture	0.20%
Carbon	45.04
Hydrogen	5.60
Oxygen	1.56
Nitrogen	0.08
Sulfur	0.14
Chlorine	45.32
Ash	2.06
	100.00%

The chlorine content of 45.32% together with an oxygen content of 1.56% bears out the above discussion as well as suggests that some of the bottles were actually a copolymer resin.

In a municipal incinerator handling 1200 tons per day of mixed household, municipal and industrial solid wastes containing an average 0.25% of polyvinyl chloride type materials (assuming an analysis such as given by Kaiser and Carotti), the theoretical maximum amount of hydrogen chloride available from the plastics will be some 1.35 tons per day. Normal household and municipal wastes, in the absence of any plastics content, contain certain chlorine bearing

constituents. These arise from such items as common
salt from the vegetable and food portions of garbage,
chlorides from the paper and paper products present
in such large amounts and, especially during the
winter and spring, chlorides from the deicers used
on highways and sidewalks.

Some investigators suggest that perhaps half of
the chlorides actually measured in a refuse analysis
originate from the nonplastics portion of the mixture.
Thus Kaiser and Carotti quote a total chlorides
figure of 0.50%, while recognizing that less than
half of this comes from any plastics source. In a
very comprehensive study made in Hamburg, Germany[3]
and reported in 1970, it was found that, on the same
basis of a chlorine content of the polyvinyl chloride
plastic as for Kaiser, namely 45%, the average chloride
content attributable to the plastics was 54.7%, with
a range from a low of 38.1% to a high of 75.5%. This
is in good accord with the U.S. data cited. Swedish
authorities also recognize that appreciable amounts
of chlorine compounds are already present in the
refuse, even in the complete absence of plastics,
as do investigators in the United Kingdom, Switzerland,
and other countries.

After an initial drying, when steam is evolved,
there is a period during which volatile organic
matter is evolved due to various processes of devola-
tilization, degradation and depolymerization. This
may start at temperatures just above 100°C and
continue up to about 300°C, during the latter part
of which period, ignition commences. In the case of
PVC, in the range of about 170°C to 300°C, the
material decomposes to give off hydrogen chloride.
At temperatures above 300°C, ignition is well advanced and the main burning and combustion occurs
from 400°C to 700°C when the gaseous products are a
mixture of water, carbon monoxide and carbon dioxide
together with the HCl from any chlorine-containing
compounds present. During this period, any inorganic
chlorides present, such as salt, will evolve HCl.

At temperatures above 600°C and up to the usual
maximum levels of temperature for most conventional
incinerators, complete oxidation takes place with the
carbon compounds present burning to carbon dioxide
and water. At these higher temperatures also there
is a certain amount of nitrogen oxides formed from
certain nitrogen compounds present in the solid wastes
and also from high temperature oxidation of the
nitrogen in the air.

As far as PVC is concerned, above about 300°C and after all the HCl has been evolved, the residue burns essentially like any other organic material. From the burning of PVC, therefore, no products other than HCl, water and carbon dioxide are evolved.

It has been reported that phosgene ($COCl_2$) is evolved in the burning of PVC, but this is completely erroneous. Well documented studies, particularly by E. Boettner *et al.* (*Journal of Polymer Science, 13*, 377-391, 1969) and L. B. Crider and M. O'Mara (American Industrial Hygiene Association meeting, May 14, 1970), show that the decomposition products do not include phosgene even at levels as low as one part per million. In a paper given by Jay,[4] a formula is derived which relates the possibility of phosgene formation on burning of the chlorinated hydrocarbons to the composition as follows:

$$\propto = \frac{\text{number of chlorine atoms} - \text{number of hydrogen atoms}}{\text{number of carbon atoms}}$$

When $\propto$ is 0.40 or more, phosgene can be formed.

For PVC:

$$\propto = \frac{1-3}{2} = \frac{-2}{2} = -1 \qquad \text{which is less than 0.40.}$$

It has also been well established that no chlorine gas is generated from the burning of PVC.[4]

What happens to the hydrogen chloride generated on incineration of solid wastes containing chlorine compounds? This is somewhat dependent on the design and operation of the particular incinerator. It would now appear that part is recombined with the ash residue of the incineration process itself. In general the residue of incineration is strongly basic in nature and will neutralize some of the hydrogen chloride present. Experiments have been performed whereby additional basic material, such as lime, has been added to the refuse before incineration; this has certainly contributed to a reduction in the amount of hydrogen chloride emitted from the incinerator. The fly ash has also been shown to be an absorber of hydrogen chloride. Again, experiments have been performed whereby lime is injected into the hot gas stream just after the main burning chamber, causing a reduction in the chlorides subsequently leaving the stack.

For those incinerators having water sprays to cool the gas and knock down the fly ash in the gas stream, there is a further reduction in chlorides in the final gas leaving the stack, while for incinerators with modern electrostatic precipitators all chlorides associated with the fly ash are removed, leaving only a residual hydrogen chloride content in the stack gas itself. Actual definitive measurements on this question are, unfortunately, not very numerous, but there are some data of value especially from Germany, Sweden, and more recently, in the Kaiser and Carotti report prepared for the Society of the Plastics Industry.

For example, Reimer and Rossi[3] in Hamburg, Germany studied this question because of plans for the beverage industry to use one-way vinyl bottles in place of glass, "with the resultant complex possibility of contaminating the atmosphere through the emission of hydrogen chloride." Their experiments were conducted in October 1969 using a 200-ton/day Von Roll incinerator and consisted first of a careful analysis of the input refuse to which was added varying amounts of PVC from 2 to 10%, and then an analysis of the chloride balance during incineration. As might be expected from theory, the investigators found a linear relationship between the amount of hydrogen chloride in the flue gas and the amount of PVC added to the refuse. At a 4% level of PVC in the refuse, the hydrogen chloride content of the flue gas was 1.7 grains per cubic foot, compared to 4.5 grains per cubic foot at a PVC concentration of 9%.

The concentration of chlorides in the fly ash was also approximately linear with PVC concentration, ranging from about 3% at 3.5% PVC content to 6.5% at 9% PVC content. In contrast the chloride content of the slag, which ranged between 0.28 and 0.37% over the whole range of PVC additions, was little affected by the PVC content of the refuse. This latter observation is consonant with the belief that at the actual temperatures of incineration, the chloride-containing materials also decompose to give gaseous hydrogen chloride which must be either reacted with the fly ash or appear in the flue gases.

The maximum long-term allowable concentration of foreign gases in the atmosphere at the earth's surface, that is where living organisms could be endangered, is given by the MIK_D value (maximum immission concentration). While there appears to be no expressed limit for hydrogen chloride in the United States, experience over many years in Europe suggests that it is of the order of 0.0003 grain per cubic foot.

Since the flue gases are strongly diluted on
entering the atmosphere, it can be calculated that
the maximum permissable concentration of hydrogen
chloride in the actual flue gas can be some 3 to
5,000 times higher or 0.88 to 1.5 grains per cubic
foot, the actual value depending on the terrain and
the height of the stack. Some U.S. authorities quote
a factor of 8,000 to 10,000 for stacks of 200 feet
in height or higher. From Swedish experiments, cal-
culation shows that to meet the 0.88 to 1.5 grains
per cubic foot figure quoted above, the actual PVC
concentration in the solid wastes would have to be
3.5 to 6%, which is many times greater than the 0.2%
that presently exists in the U.S. average waste
composition.

In a paper by Rasch[5] dealing specifically with
the role of sodium chloride in high temperature
corrosion, reference is made to the literature and
especially to work at BASF. The fact that any
sodium chloride present in the refuse can have
sufficiently high vapor pressure at furnace temperatures to take part in corrosion reactions is brought
out, but otherwise the paper supports the general
conclusions discussed earlier. It can, therefore,
again be stated that to avoid corrosion in incinerators employing heat exchangers for waste heat
recovery, it is important to operate at temperatures
above the point at which condensation can occur,
below the temperature at which melting of the fly
ash takes place, and in a strongly oxidizing atmosphere. A suitable temperature range in the heat
exchangers would be between 150 and 450°C, and
preferably between 150 and 350°C.

It has sometimes been stated that the incineration of plastics wastes containing nitrogen (for
example, the nylons, the ABS materials or other
more recent candidates for bottles) will give rise
to the formation of nitrogen oxides, which will
enhance corrosion of incinerators and be detrimental
to air quality. A specific study of this question
was made, and no support was found for any contention
that nitrogen oxides are liberated as a direct result
of the incineration of such plastics wastes.[1]

In the normal operation of an incinerator,
nitrogen oxides are present in the flue gases to a
very minor extent. This is shown in the report by
Kaiser and Carotti,[2] and it is of significance that
even in the case of burning a refuse containing 4%
of a polyurethane material with some 6% nitrogen
content, the amount of nitrogen oxides found in the

flue gas was not significantly higher than in the
samples taken of flue gas from refuse burned without
additional polyurethane. It would now appear reasonable to state that any increase in nitrogen oxides
reported when nitrogen-containing plastics are
incinerated is probably due to the reaction between
the nitrogen and oxygen present in the air in the
furnace taking place at a somewhat higher temperature
than normal. This is supported by the data for
incineration of wastes containing 2 and 4% additional
polyethylene where the nitrogen oxides produced in
the absence of additional polyethylene was 28.6 ppm
versus 40.1 and 39.9 ppm for 2 and 4% additions
respectively, and the polyurethane data where the
corresponding figures are 32.2, 39.6 and 39.5 ppm.

Liu, Theoclitus and Dervay[6] studied the incineration of high BTU solid waste such as plastics on
controlled air principles. These principles are
delineated in a companion paper by Theoclitus, Liu
and Dervay.[7]

Controlled air incinerators can be operated
continuously, intermittently or in batch using
various temperature and air distribution controls.
Because of the flexibility in design, new incinerators capable of burning 100% plastics or other high
heating value waste have been evolved and marketed.
Such incinerators not only can serve isolated industrial establishments and institutions for on-site
waste disposal, but also can be grouped in multiple
units to economically serve commercial centers,
suburbs, and small communities for commercial and/or
municipal waste disposal. Due to their modular
sizes, several groups of these multiple units can
be strategically located around the community to
minimize high hauling costs.[7]

There are two main approaches for the combustion
of polymers:

1. *Single-Stage Combustion.* This includes direct
   burning both in the open atmosphere or in incinerators
   of high excess air design.
2. *Two-Stage Combustion.* This is accomplished by combustion of the waste in an oxygen-deficient
   environment and complete oxidation of the total
   resultant gaseous combustibles in a separate chamber
   at controlled excess air.

The combustion process of the high BTU waste in
the chamber of the standard controlled air incinerator
proceeds in stages. Following the start-up, aerobic

burning and anaerobic pyrolysis occur simultaneously. With the assumption of perfect utilization of the air in the chamber, the fixed amount of chamber air sustains the combustion of a fixed quantity of high BTU solid waste, except at the beginning and at the end of the burn. In general, the heat released from the combustion of the waste is constant throughout the burning cycle. Steady as the combustion is, the heat loss during the burn is transient in nature.[6] It includes:

1. Heat absorbed by the incinerator chamber refractory;
2. Heat loss by heating the rest of the waste in chamber;
3. Heat loss by volatilizing or gasifying part of the waste; and
4. Heat carried out by chamber gases as latent heat of the volatiles and combustion products.

In operation, a continuous air modulation system is installed to modulate the chamber air and is controlled by a thermocouple located in the stack. When the stack temperature increases, the air supply decreases. A similar modulation system for the afterburner fuel is also installed. The burner is modulated between maximum and minimum fuel limits. When the stack temperature decreases, the afterburner heat generation increases. The controlling points for the afterburner may be set lower than that for the chamber air to economize fuel consumption whenever possible. The burner capacity may be so selected that it is capable of preheating the stack to the same temperature level as required for clean incineration of the high BTU waste.

At the start-up, the stack temperature is preheated to the desired level, and start-up smoke is eliminated. As the stack temperature rises rapidly and reaches the afterburner set point, the afterburner then starts to modulate. When the stack temperature passes the chamber air set point, the modulating chamber air regulates the burning rate. Hence, the volatilization rate in the chamber is kept at a steady level so that the volatiles or gaseous combustibles can be completely incinerated in the stack where the stack combustion air is supplied at a constant rate. During the burndown stage, the afterburner is operated at maximum capacity to keep the stack temperature high enough for clean burning. The burner is then turned off by a timing device at a predetermined time which may be established by field experience.[6]

Applications for batch-burning high BTU incinerators which have been studied and reported include
polystyrene foam waste generated from fabrication
processes. The material has a heating value of
18,120 BTU/pound and is essentially styrene polymer
with very minor amounts of organic additives. At a
burning rate of 260 pounds per hour, the stack
emission level, including condensibles, is only
0.087 grain per standard cubic foot at 12% CO_2 or
0.077 grain per standard cubic foot at 50% excess
air.[6]

Staudinger, in the S.C.I. report Number 35[8]
covers many of the aspects of the incineration of
plastics in the United Kingdom. He states that it
has been established that present-day domestic and
trade refuse in the United Kingdom contains between
1 and 2% of plastics (0.05 to 0.1% being PVC),
mainly of packaging origin, and it has become evident that any progress towards better living standards and any increase of material affluence will
increase the quantity and will affect the quality
of wastes. Plastics from one of many classes of
materials which respond strongly to the affluence
factor, and a continuous increase of plastics can
be expected in the refuse of the future. It has
also been shown that by 1980 plastics could well
constitute one-twentieth of all municipal refuse.
It may therefore be useful to consider here the
behavior of the various plastics materials towards
heat, burning and oxidation.

The whole spectrum of plastics can be divided
into two classes, the thermoplastics and the thermosetting materials. The thermoplastic materials, as
the name implies, soften, deform and melt when
heated, whereas the thermosetting plastics, once
converted into their end-use type of article, are
no longer deformed when heated. On this basis alone
their burning characteristics will be different.

In the case of polyethylene, polypropylene,
polystyrene and other polyolefins, the elements
connected to the carbon atoms consist only of hydrogen or groupings of carbon atoms, again linked to
hydrogen or groupings of carbon atoms, again linked
to hydrogen atoms, so that these plastics can be
described as hydrocarbons. When these materials are
burned with an excess of air, the end products are
carbon dioxide and water, although in the process of
thermal oxidation they first undergo a breakdown
(depolymerisation and degradation) into volatile
products. This holds also for polymers containing
oxygen atoms as well as carbon and hydrogen,

irrespective of whether the oxygen is present in the form of ether, carbonyl, carboxyl or hydroxyl groups. However, the situation is more complex in the case of PVC in which some of the hydrogen atoms are replaced by chlorine atoms. It is this replacement which is responsible on the one hand for the many valuable properties of the polymer, but on the other hand it is also the cause of possible disadvantages in burning it.[8]

A characteristic of plastics is their high calorific value which in many instances equals that of high-grade fuel oil. However, in polymers where the carbon-hydrogen content is diluted with other elements, such as chlorine, nitrogen or oxygen as in PVC, the polyamides or polyacrylates for example, the calorific value is lower as can be seen in Table 22.1.

Table 22.1

Calorific Value of Plastics[8]

Polymer	BTU/lb	Kg cal/kg
Polyethylene	18,000	10,000
Polystyrene	19,500	10,800
PVC	9,500	5,250
Polymethyl methacrylate	14,000	7,750
Polyamides	13,000-15,000	7,200-8,200
Polyesters	13,000	7,200

The S.C.I. report indicates that when large quantities of plastics waste from industrial or trade sources are added to the domestic refuse or when industrial waste with a high plastics content has to be incinerated and all design factors such as the higher BTU value, the air supply requirements, combustion time, burn-out temperature, etc., have been taken into account for all types of incinerators, the preference in the United Kingdom may well be the rotating drum principle. There are two reasons for such a choice: first, the greater degree of blending and mixing of the refuse bed during the whole of the combustion process; and second, the fact that any pool of molten plastics which may have formed under the refuse bed is continuously being exposed to the combustion air by the rotation of the drum. Some of the industrial waste incinerators used by the chemical industry are based on this principle. Although

plastics are but one of the very many materials which
make up industrial chemical waste, brief reference
is made here to information in the literature where
the industrial waste contains or consists of plastics.

In general one can say that wastes from the industries which produce the chemical raw materials
and the polymers for the plastics industry are of
considerably greater complexity in their composition
and consistency than urban refuse, in that liquids
(inflammable, toxic, harmless) sticky, gooey residues,
organic and inorganic solids, disused packaging
materials (barrels) and general factory refuse may
be present in proportions which vary from day to day.
In addition the heat release value (BTU), the rate
of burning, the density, and the oxygen demands vary
within wide limits, so that incineration requires
specially designed equipment and great skill in
operation.

The conclusion which one can draw from this information is that plastics as industrial waste can
be successfully disposed of by incineration provided
that the design of the installation and its operation
take into account the peculiarities of the waste.
Furthermore, it appears possible and in some cases
perhaps economically advantageous to include the
necessary design provisions when planning new
municipal refuse incinerators for the addition of
industrial plastics waste from neighboring industries
to the domestic refuse.[8]

Specially designed incinerators are available
for the burning of waste PVC, where its recovery as
such is neither desired nor possible, but where
circumstances make it worthwhile to recover at least
the HCl for reuse.[8]

In the matter of air pollution, Staudinger[8]
points out that even without the presence of plastics
in the refuse, the incineration effluent gases contain undesirable pollutants, and the presence of
plastics such as PVC, PTFE or polyurethane will
aggravate this situation. However, for the foreseeable future, there is no likelihood that hydrofluoric acid from fluorine-containing plastics will
become a problem, whereas PVC, if present in
sufficiently large quantities in the refuse, could
in certain cases create conditions which would make
it necessary to apply practical remedies now.
Therefore, the important question here is how much
PVC could be tolerated in the refuse so that the
HCl emission does not exceed the level of 0.2 grain
per cubic foot (or about 450 ppm), a level which

has been recommended as a maximum for the U.K.
Assuming that all the chlorine in the PVC is converted into hydrochloric acid, and assuming also
that no removal of hydrochloric acid occurs through
absorption in the fly-ash, conversion into chloride
of iron, or by reaction with alkaline materials in
the refuse bed, the refuse would have to contain
0.5% PVC to reach this level. This is about five
times the PVC concentration in today's refuse, and
if only the immediate future were to be considered,
say the next five years, there would be nothing to
worry about. However, there will come a time when
the PVC content will reach levels of 1 to 2% and
the question arises whether this could lead to conditions where the correspondingly greater amounts
of HCl escaping into the atmosphere would constitute
hazards to health and environment.

Such a situation might arise by 1980 in the U.K.
when the amount of PVC wastes from all sources will
be of the order of 200,000 to 250,000 tons. If it
is assumed that by that time 20% of the refuse will
be dealt with by incineration, which is very likely
to happen, an equivalent of 25,000 to 30,000 tons
of HCl could be released into the atmosphere. If
this is compared with the major air pollution of-
fenders, carbon monoxide and sulphur dioxide, which
are considerably more injurious to health, and which
by their annual production in immense quantities
(several million tons) would mask any effects which
HCl can have on health and environment, one comes
to the conclusion that air pollution by HCl will not
be a serious problem. To further support this statement, one must not overlook the fact that the
hydrochloric acid quantity found in the flue gases
very rarely corresponds to the amount which could
theoretically be given off by the decomposition of
PVC. Practical tests to determine the hydrochloric
acid content in the flue gas and to relate this to
the amount of PVC in the refuse usually show such a
deficiency. Evidence including some already cited
is now available from several sources. For example,
in Sweden[9] incineration experiments were carried out
with refuse which was enriched with PVC to 2.2%, *i.e.*,
seven times the normal PVC content. With no special
steps to remove the HCl from the combustion gases,
only 60 to 65% of the theoretical quantity appeared
in the flue gases. When at the same time lime was
added to the refuse, the HCl escape in the flue gases
was still further reduced to about 45% of the
theoretical quantity. The same report refers to

research work in which the flue gases were used to
dry sewage sludge, when it was found that 80% of the
acidic components, *i.e.*, oxides of sulphur and HCl,
were removed by the alkalinity of the sludge.

Further reports on practical experience from
Sweden[10] mention the results of hydrochloric acid
emission analyses of four incineration units, two
of which were linked to waste-heat boilers, whereas
the other two had apparently no such indirect cooling
of the flue gases. The former gave 62 and 25 ppm
hydrochloric acid while the incinerators without
boilers gave 135 and 105 ppm hydrochloric acid. The
PVC content in the waste has been of the order of
0.1 to 0.2%, which theoretically should give 100 to
200 ppm hydrochloric acid. Here again the practical
results are below the corresponding theoretical
value.

The results of a more detailed experiment have
been reported in which an urban refuse, enriched
with PVC to 5%, was incinerated. After sufficient
time had elapsed for conditions to stabilize, careful chlorine analysis on the flue gas, fly-ash, slag
and quench water were carried out with the following
results: flue gas, 51% of the theoretical value;
fly-ash, 4.6%; slag-ash, 4.6%; quench water, 0.3%.
In other words about 40% of the theoretical quantity
of HCl cannot be accounted for, but the important
point here is that only half of the possible quantity
of HCl appears in the effluent gas.

The conclusion which one can draw from these
observations is that in modern incinerators with
waste-heat boilers or high-pressure steam generation
and with the appropriate flue gas purification
equipment, such as dust arrestors, multicyclones,
filters or electrostatic precipitators, there is a
likelihood that the actual hydrochloric acid emission
will always be below the theoretical value. However,
this should not be used as a basis for future situations, and unless conditions inside the incinerator
can be created which prevent the escape of hydrochloric
acid with the flue gas, suitable equipment must be
installed, such as wet-scrubbers, to eliminate or
substantially reduce this particular pollutant as
well as the much more noxious oxides of sulphur.[8]

From Japan, a specialized case of plastic waste
incineration is reported.[8] Atactic polypropylene,
which accumulates in regular quantities from the
large-scale manufacture of polypropylene, is used
as the sole fuel in a steam-raising incinerator with
a specially designed burner unit. This is perhaps

a unique case, but it demonstrates that the high heat content of plastics can be utilized to advantage.

Refuse in Japan has a far higher moisture content and lower calorific value (lower heating value: 2000 to 5100 BTU/pound, 40 to 70% moisture content) than that of Europe or America (lower heating value: 4000 to 10,000 BTU/pound, 10 to 45% moisture content). The refuse in Japan has a higher percentage of garbage and a lower quantity of paper plus more than three times the plastics content in raw refuse (see Table 22.2).

Table 22.2

Annual Change in Plastic Content of Raw Refuse[1]

Year	Change in Weight Percent
1963	2.7
1964	3.4
1965	4.2
1968	9.1

The amount of vinyl chloride resins contained in raw refuse can be estimated based on the fact that approximately 35% of all plastic products in Japan are vinyl chloride resins. Moreover, from the table it can be seen that the amount of plastic waste contained in refuse is increasing annually, making corrosion due to the presence of HCl gas an ever-increasing problem.

By examining the values for refuse in Europe and America, it can be readily seen that the refuse in Japan has a much higher moisture content and lower calorific value. Although refuse is a mixture of various kinds of material, the composition and calorific value of the combustible content in the refuse in Japan are comparatively constant (see Table 22.3).

It is well known that the fuel combustion rate is affected by the calorific value of the fuel, temperature of air used for combustion, type of grate, etc. From calculations it can be concluded that Japanese incinerators designed for processing refuse which has a low heating value of the order of 2700 BTU/pound must have a grate area of almost twice

Table 22.3

*Chemical Composition and Calorific Value
of Combustible Content*

Carbon	42 - 56%
Hydrogen	5 - 7%
Oxygen	40 - 48%
Sulfur	0.3 - 0.9%
Nitrogen	0.6 - 1.9%
Higher Heating Value	2500 - 3700 BTU/Pound

that of incinerators of equivalent capacity used in Europe and America.

As in Europe and America, strenuous effort has recently been made to improve and expand environmental sanitation equipment and facilities in Japan. Matsumoto, Asukata and Kawashima[11] describe briefly the history and development of the Japanese refuse incinerator facilities up to the present data. Also discussed are incinerator variations dealing with the higher moisture content and lower calorific value of Japanese refuse, several Japanese incinerator plants, and other related topics.

To cope with ever increasing requirements for adequate environmental sanitation facilities, the Japanese government passed emergency legislation in 1963 calling for environmental facilities. A 5-year plan for the construction of incinerators based upon the new law was contemplated. Engineering studies were conducted on refuse incineration, and incinerators used in Europe and America were investigated. Test equipment was developed, and the design and production of continuous feed, mechanical incinerators was commenced. In the course of the design and construction, consideration was given to public health and sanitation to prevent public hazards and nuisances.[11]

The first equipment of this type was installed in the Osaka Sumiyoshi plant (150 tons/day x 3 units). The Sagamihara city plant (90 tons/day x 2 units), and Tsurumi plant, Yokohama (150 tons/day x 3 units) were completed at about the same time. These plant designs were based upon the different conceptions of combustion and incineration. Following the construction of these plants, other plants were then built using various types of incinerators.

In Japan, incineration service is provided either
directly or indirectly by public service organizations,
and, in view of the recent public interest and growing
concern about prevention of public hazards and nuisance
stringent measures have been taken to provide assured
protection. For example, dust collectors have been
installed, and consideration has been given to the
elimination of unpleasant odors, prevention of water
pollution, and noise control. Although proper refuse
incineration requires a high degree of technical
knowledge to cope with these conditions, only a few
engineers have concerned themselves with this subject.
As difficulties were encountered in the development
of suitable refuse incineration plants, there was a
tendency to establish standard specifications with
respect to such facilities. Public interest on this
subject was aroused, and appeals from local public
service organizations began to come in to the various
technical institutions. Thus, numerous studies have
been conducted on analysis of refuse, problems of
public hazards and nuisances, and theories of drying
and combustion. This has brought about considerable
discussion and joint efforts by the Ministry of
Health and Welfare, university scholars, local public
service organizations, manufacturers, and has resulted
in the development of a Standard for Incineration
Facilities which was set forth in 1966. These maintenance and control standards for refuse incineration
facilities have since provided a stringent policy
guidance for incinerator construction as well as
effective guidance to manufacturers of incinerator
equipment with respect to future development of
incinerating equipment and facilities.[11]

As a result of the growing interest in this subject, new mechanical incinerator plants have sprung
up in one city after another. The government
responded to the concern for improved sanitation
facilities by developing a new 5-year plan that
begin in 1967. A brief summary of this plan is
given below.

Number of cities, towns and villages requiring refuse incineration facilities	1,273
Refuse to be processed	61,650 tons/day
Processing capacity at the end of 1966	27,685 tons/day
Additional amount to be processed during and after 1967	33,965 tons/day

Breakdown

Amount to be processed by
  continuous mechanical furnaces   19,530 tons/day
Amount to be processed by
  batch-fed furnaces   14,435 tons/day

(Data obtained from a study on Long-Term Plan for
Development of Facilities for Daily Living, published
by the Environmental Sanitation Division of the
Ministry of Health and Welfare.)

REFERENCES

1. Warner, A. J., C. H. Parker, and B. Baum. "Solid Waste
   Management of Plastics," study prepared for Manufacturing
   Chemists Association (December, 1970).
2. Kaiser, E. R. and A. A. Carotti. "Municipal Incineration
   of Refuse with 2% and 4% Additions of Four Plastics:
   Polyethylene, Polyurethane, Polystyrene and Polyvinyl
   Chloride," a report to the Society of the Plastics
   Industry (SPI) (June, 1971).
3. Reimer, H. and T. Rossi. *Mull and Abfalle,* pp. 1971-1974
   (March, 1970).
4. Jay, J. *Chemie et Industri,* Vol. 92, pp. 533-537 (1964).
5. Rasch, R. *Energie,* Vol. 23, Number 2 (February, 1971).
6. Liu, H., G. Theoclitus, and J. R. Dervay. "Incineration
   of High BTU Solid Waste and its Applications," paper
   prepared for 1972 National Incinerator Conference (ASME)
   at New York (June, 1972).
7. Theoclitus, G., H. Liu, and J. R. Dervay. "Concept and
   Behavior of the Controlled Air Incinerator," paper prepared for 1972 National Incinerator Conference (ASME)
   at New York (June, 1972).
8. Staudinger, J. J. P. "Plastics Waste and Litter," S.C.I.
   Monograph Number 35, published in England by Staples
   Printers Ltd.
9. "Plastics and the Environment," Report No. 160 of the
   Academy of Engineering Science, Stockholm (1969).
10. Heidenstam, G. V. "Burning and Incineration in Sweden,"
    paper presented at Air Pollution Association Meeting at
    New York (June, 1969).
11. Matsumoto, K., R. Asukata, and T. Kawashima. "The Practice
    of Refuse Incineration in Japan," paper prepared for 1968
    National Incinerator Conference (ASME) at New York
    (May, 1968).

CHAPTER 23

PLASTICS IN SANITARY LANDFILL

Since plastics are for the most part inert and
will not decompose to give off hazardous or distasteful odors or gases, landfill methods can prove to be
very efficient and practical solutions to the problems
of disposal in the community or in industry where
significant quantities of plastics materials are
found in solid wastes.

In general one simple overriding statement can
be made: Plastics materials in landfill act essentially as inert material. In regard to changes in
composition with time under the influence of soil
organisms, rodents or other animal life, there
appear to be none at all comparable to the decomposition of vegetable matter or paper and paper
products, nor do the putrescible components of the
refuse while undergoing their own decomposition have
any effect on the plastics materials. It is true
that certain organic materials added to the base
resins to assist in the processing of the plastics
compounds may be susceptible to attack by certain
micro-organisms, or may be leached out over a period
of time. However, there is no evidence that this
occurs other than to a minor extent, nor are these
processes and their products different from the
status of a landfill in the absence of plastics.
The unique characteristics of plastics, which have
made them so widely accepted as corrosion resistant
and environmentally inactive, carry over to when they
are buried. It has been said that in digging up old
landfills after many years of service, items made of
plastics have been recovered in practically identical
nature to the original.[1]

The basic effect of plastics on landfill is therefore in the area of density. Still, as has been

pointed out before, the bulk of the plastics entering
the solid waste stream originate from the packaging
industry where they are used as films, overwraps,
bottles, jars, tubes and other forms of containers.
In general, they are not brittle and therefore do
not readily break up on impact to smaller fragments.
Even under moderately heavy loads, they will not
easily disintegrate, but rather deform and tend to
return to their original shape when the load is removed. In a sanitary landfill, therefore, plastics
tend to confer a somewhat lower finished density to
the landfill. Thus, improper landfill practices
whereby excessively high concentrations of plastics
materials are placed in one spot (for example, a
truckload of scrap bottles just dumped with no attempt made to spread them over a large area), could
lead to uneven landfill settling and final compacted
properties, and could reduce the bearing load for
subsequent roads or structures to be placed on the
completed site.

Another potential problem in sanitary landfill
is the trapping of the gases formed during the decomposition of the landfill refuse. No way has yet
been found to prevent production of methane gas or
odorous gases such as hydrogen sulfide. A proper
sanitary landfill has provision for the venting of
such gases formed in a safe manner and avoids trapping of these gases in isolated pockets. Polymeric
materials are not known to contribute to such gas
production.

In England and Western Europe, some confusion
exists in the terminology of solid waste disposal
when such terms as "open dump," "landfill" and
"sanitary landfill" are employed. In the United
States these terms are reasonably defined and
understood, but such terms are not generally employed
in Europe. In the United Kingdom, for example, the
term "controlled tipping" is analogous to the United
States "sanitary landfill" and has been practiced
for some fifty years. It is the least costly and
the most widely practiced method for the disposal
of domestic garbage. It consists of spreading the
refuse with the obligatory layer of soil in a suitable location, where exposure to the elements causes
rusting, oxidation, decay, and microbial breakdown
of certain of the organic constituents such that
compaction eventually takes place with a loss of
individual identity.

Staudinger, in the S.C.I. report,[2] states that
the actual process of transferring the refuse from
the vehicle into the landfill has certain disadvantages

in that any light-weight constituents (paper, plastics films, thin-walled cups and the like) can drift or are blown about. Windy conditions will aggravate the situation, and very strong winds can create a "blizzard" of such materials. Although this constitutes a considerable nuisance, for which plastics must take their share of the blame, it does not alter the fact that controlled tipping is still very widely used. Once the refuse has settled and has become stabilized in its density, drainage and water retention characteristics, such tipping areas can be restored to useful land. It is here that the presence of large quantities of plastics waste in the refuse may cause objections in that it can lead to conditions in which the processes of settling and stabilization may be retarded or even prevented by the creation of cavities, air pockets and loosely packed aggregates. Drainage too may be affected by plastics films forming barriers in the refuse profile. Plastics bottles and other types of plastics containers, which do not decay, degrade or rot even after prolonged soil burial, are not always easily squashed, and only those in the lower strata of the refuse may become distorted, flattened or compressed by the overlying weight. Those with less weight of over-burden are likely to remain intact and thus contribute to the "springiness" of the top layer. At the present time, with refuse containing only 1 to 2% plastics, of which less than half appears as undamaged hollow-ware, their effect is not likely to be serious. However, once the concentration reaches a level of say 5 to 6% or more in the tipped refuse, recovered land from such tipping areas may not be suitable for supporting roads, light building constructions or any other load-bearing use, and may be usable only for landscaping and vegetation cover. It is known that growth and root formation of plants, shrubs, and trees remain unaffected when coming into intimate contact with plastics such as polyolefins, polystyrenes, rigid PVC, polyesters, urea-formaldehyde resins, synthetic rubber, etc. Concentrations of that order are likely to occur in the U.K. once the per capita consumption of plastics reaches a level of 150 to 180 pounds per year, corresponding to an annual consumption of plastics of about 4 to 5 million tons which by analogy with the trends in the U.S.A., Japan, Sweden and Germany, may be expected between the years 1980 to 1990. Whether by that time sufficient area and volume of tipping sites will still be available in reasonable proximity to the refuse-producing centers

depends very much on local situations. Under adverse
conditions other disposal methods, less demanding on
tip volume, will have to be adopted. Or as an interim
measure it might be necessary, in order to prolong
the life of the available disposal sites, to reduce
the refuse to the smallest practical volume.[2]

CHANGE IN PHYSICAL STATE

Volume reduction can be achieved by compaction
(off-site or on-site) or by pulverization, but in
either case it calls for an additional mechanical
treatment which adds to the cost of the disposal.
However, on many occasions this added financial burden may be less than that from the use of more remote
disposal sites involving raised transport cost. In
addition, pulverization or compaction results in a
much denser refuse with better settling and stabilization characteristics, thus allowing the return of
the area to useful purposes in a much shorter time.
Whatever process is used to compress the refuse
(on-site or off-site compaction or pulverization),
the result will be a reduction, if not a complete
elimination, of the deleterious effects plastics may
have on the tip characteristics. Pulverization can
reduce the refuse to half or even less of its original
volume, and on-site compacting of the freshly deposited
layer of refuse by tracked vehicles such as bulldozers
used for spreading the delivered refuse is claimed
to result also in a volume reduction down to half.
In the case of pulverization the presence of plastics
materials in comminuted form will no longer contribute
to the springiness of the deposit and will have little
detrimental effect on other tip characteristics such
as settling speed, levelling, stabilization, water
retention, and drainage.
On-site compaction, apart from reducing the
springiness of the refuse bed, tends to create
horizontal laminae mainly with large items of waste
such as cardboard boxes, wooden cases, plastics
cannisters, and the like. These waste items are
flattened, and if such horizontal layers are not
interspersed with a multitude of small refuse components, and even squashed plastics bottles, tins
and broken glass, surface water drainage may become
unsatisfactory. However, such simple methods of
compaction produce a reduction in the free volume of
air inside the refuse, which in turn helps to prevent odor development, rodent intrusion, and

fermentation. For more efficient on-site compaction, traction vehicles have been developed which rely on heavy wheels instead of tracks. The rims of these cast iron wheels have specially designed ribs and other protrusions capable of breaking and crushing even large waste items. In this way a high degree of compaction is achieved without a tendency toward stratification of the daily deposited refuse. Whatever the merits of this treatment, it makes possible a larger volume acceptance of the tip with the dense refuse contributing to stable conditions in a shorter period of time than can be realized with uncompacted refuse containing plastics.[2]

Staudinger, in the S.C.I. report, states that any treatment which converts the refuse into a more particulate state, irrespective of whether this is done by pulverization, shredding, crushing, or shearing, is advantageous to any subsequent disposal process. Hence, such refuse treatments appear attractive not only for the reasons mentioned above such as increase in tip capacity, prevention of odor development and rodent infestation, but also on account of the usefulness of the pulverized refuse *per se*. It has been suggested that pulverized refuse needs little or no soil covering and can even be used instead of soil in cases where the latter is scarce and where untreated refuse has to be covered. Pulverized refuse is most useful also where it is intended for the restoration of spoiled lands, for landscape improvements and for land elevations, *i.e.*, in all cases where for economic, aesthetic, safety or other reasons the contour of the land surface requires changes. In addition, many of the pulverizing, grinding and shredding processes involve a partial separation of some components from the refuse, *i.e.*, ferrous metals, rubbery materials, soft plastics or plastics bottles, the removal of which contributes to an improvement of the refuse quality. Once certain components have been separated, the possibility of salvaging them for recovery and reuse exists, and there is a tendency to reduce the financial burden of the pulverization process by the proceeds of their sale. In the case of metals it might be worthwhile, but with plastics the salvaged material would scarcely find a market and thus would have little or no financial value.

Machinery for refuse pulverization is based on attrition, impact, shearing or crushing actions, and a great variety of types based on any one of these principles is available both here and abroad. The

question of particular interest here is what happens
to the plastics waste during these processes, because
this aspect will assume considerable importance once
the plastics content in the domestic refuse increases
greatly over and above its present level to 1 to 2%.

One can broadly classify such machinery into
rotating drums (similar in type and action to ball
mills), centripedal hammer-mills, differentially
rotating rasps, and multi-roll crushers. Although
such machines are extensively used in industry for
a great variety of purposes, *i.e.*, grinding of ores,
minerals, solid chemicals, crops, and foodstuffs,
some have been modified in one way or another to be
able to cope with such a heterogeneous mass as refuse.

The response of plastics to the various actions
in such machinery depends not only on their properties, such as hardness, impact resistance, flexibility,
resilience and rebound characteristics, but also on
the form in which they present themselves. For
example, the rotating drum pulverizers, with or
without internal counter-rotating beater arms, have
little disintegrating action on plastics in general,
although hard or brittle or thin-walled articles
such as gramophone records, foamed packaging trays
and the like become cracked and broken by the
churning and tumbling action.

The S.C.I. report[2] further states that, anything
that cannot pass through the holes in the screening
section of the drum is discharged at the outlet end
of the drum, and with it the greater portion of the
plastics content together with fabrics, lumps of
agglomerated paper, glass, etc. Even so, one finds
small moldings or fragments of moldings and some
remains of film in the treated refuse, but their
amount is so small that their presence can be of no
consequence to any of the subsequent disposal
processes.

The action of hammer-mills depends primarily on
impact, but judging from what happens to plastics
in such mills, to some extent also on a shearing and
shredding action. The extent to which plastics are
either disintegrated or not accepted and ejected
depends on many design factors and operating conditions, not only on the nature of the plastics
materials. Nearly all types of mills have provisions
for the rejection of oversized items (tires or drums)
and ungrindable materials (steel pipes, metal cans
and tins). Among these ungrindables are also plastics
and natural or synthetic rubber articles, where impact
fracture is prevented by their resilience, elasticity,

softness or low volume density. For example, polystyrene moldings, including foams, phenolic and
amino resin moldings and other hard plastics, are
readily shattered and broken up in the mill, while
soft articles such as plasticized PVC or low-density
polyethylene hollowware are thrown out by ballistic
action. However, no clear-cut classification between
"soft" and "hard" is produced by this mill action.

Separation or extraction of refuse components,
if not part of the actual grinding process, can be
carried out before the refuse enters the mill or on
the pulverized refuse itself. The pre-mill separation is often done manually on a picking belt and
concerns mainly oversized items, nonferrous metals
and rags, with ferrous components being extracted
magnetically. The separation at this stage does not
involve plastics unless their removal is solely
necessitated by their size.

One type of hammer mill incorporates provisions
for the ballistic separation and removal of the
ungrindables by having a tower mounted over the
hammers and allowing the refuse to "float" above
the rotor until it is either dealt with by the
hammers and passes through the grate at the bottom
of the mill or is thrown upwards into the reject
tower and discharged via a side chute for subsequent
separate treatment. This is of particular interest
from the plastics point of view.

For example, plastics such as phenolics, polystyrene, and the more rigid polyolefins (high-density
polyethylene or polypropylene) are broken up into
small pieces, together with torn strips of plastics
film (polyolefin, not larger than the palm of a hand)
and shreds of light fabrics (stockings). These
otherwise not easily grindable materials may have
become entangled with the grindable components of
the refuse, and the tearing, shearing and rubbing
action between the hammers and the casing is certainly responsible for the torn strips, shreds and
small pieces of these flimsy and filmy materials.
On the other hand the rejected material contained a
moderate amount of rubber and plastics articles such
as playballs, pieces of garden hose, parts of rubber
boots (overshoes), soft toys, bottles and many other
plastics articles of unidentifiable origin.

Experience so far indicates that pulverization
machinery based on the hammer mill principle could
readily cope with refuse containing much higher
concentrations of plastics than that occurring in
today's domestic and trade refuse. But whether it

is desirable to have a high degree of separation of
plastics in the course of such an operation is questionable because their subsequent disposal might
require special treatment. For some uses of pulverized refuse there will be no need for the total
removal of the plastics materials provided they are
present in fragmented and comminuted form like the
rest of the refuse. A fairly wide selection of such
machinery of various designs, sizes and capacities
is available even among those which have been modified, adapted, or specially designed for the handling
of domestic refuse and other wastes.

Another type of machine for disintegrating refuse
and separating out the ungrindables and large-sized
components are the *raspers*, in which refuse is forced
against protruding pins supported on a perforated
plate. The treated refuse falls through the perfor-
ations onto a lower level and is swept by rotating
arms toward the outlet. All materials which do not
respond to the crushing pressure, and thus do not
break up, are rejected through a separate outlet.
The amount of rejects is fairly high and probably
contains a large proportion of the plastics originally
in the feed.

Where circumstances demand or financial considerations allow, the use of two or more pulverizers in
series of the same or of different types is practiced,
particularly where bulky waste has to be treated on
the same site as urban refuse. Any plastics components which are part and parcel of the bulky waste,
such as happens to be the case in freezers, refrigerators, radio and television sets, furniture and the
like, are broken up together with the bulky waste
items into still fairly large fragments which, after
magnetic separation of the ferrous metals, are then
subjected to further pulverization (or separation)
in any of the above-mentioned grinders.

The technology of converting solid waste, large
and small, into particulate form by grinding, crushing or shredding has made considerable progress and,
were it not for the additional cost, pulverization
alone on account of the plastics content in the refuse
would be a most useful pre-treatment of wastes for
subsequent disposal. The heterogeneous composition
of solid wastes makes such a treatment desirable,
because it would eliminate or materially reduce the
objectionable effects attributed to plastics in
sanitary landfills, land reclamations and incineration, particularly in fluid bed or other forms of
suspended combustion.[2]

In Germany, several grinding operations were observed.[3] The advantages of grinding are a volume reduction and a reduction or elimination of vector and nuisance problems that are associated with tipping operations not using cover material. The operations observed were new, the oldest having been in operation for about two years prior to 1967.

The ballistic separation principle resulted in the Gondard mill being the predominant equipment. The mills without ballistic separation generally would require some refuse preprocessing that increases both capital and operating costs. The Gondard mill uses a 140-hp motor to drive a single rotor having 4 rows of 12 hammers at a speed of about 1,000 rpm. Refuse is fed into the mill at a variable distance above the rotor, and as the material falls on the rotor, hard, resilient and heavy objects are propelled into a vertical trajectory of about 20 feet to a point where a deflection plate directs the material into a storage compartment. Soft, inelastic material is beaten through the mill grate by the rotor.

Experience has found that the refuse moisture content affects the ability to grind the material. If the refuse is too wet, it can plug the rotor and stop the operation; if the material is too dry, the mill production is reduced because the material cannot pass through the grate. At one installation, a fire hose was used to wet the material as required. Nylon stockings were found to be difficult to grind in all mills. The stockings would wrap about the rotor and eventually would have to be cut away. It was found that about 600 tons of refuse could be processed before the hammers would have to be reversed or replaced. There was a shortage of operational cost data, but the indications were that the grinding costs were between those for controlled tipping and incineration or composting (about $2.50 to $3.00 per ton). It was interesting to find that no reliable data or reports were available documenting the actual volume reduction achieved by the grinding-tipping process. It would seem desirable to have such information before a large investment was made in the necessary equipment. Also, since the operations are relatively new, no data are available as to expected settling rates of this type of fill material.

In the villages and smaller towns of many of the European countries, small, uncontrolled "tips" (or open dumps) are still used and afford the same unsightly nuisance as in the U.S. Hart[4] reports on visits by a U.S. study team and states that large controlled landfills (in German the term is "geordnete deponie") were visited in Berlin. The solid waste disposal problem in West Berlin is extremely interesting. Although an appraisal of the Berlin situation is not yet directly applicable to anything facing American metropolises, it may be suggestive of the future when communities cannot export waste to the surrounding countryside--because there will be no countryside. West Berlin is an island of 185 square miles (roughly triangular with base and altitude of 20 miles) within the heart of politically opposite East Germany. There is essentially no trade between West Berlin and either East Berlin or East Germany. Almost all food and goods of the viable, modern, western-oriented city of 2.2 million inhabitants must be shipped in from West Germany. The cost of shipping out the wastes is obviously prohibitive, so the domestic, commercial and industrial refuse and the construction-demolition debris must be sequestered within the 185 square miles. There are presently five burial sites for refuse. Several of them began as "Trummerberge" or rubble mountains during the early postwar days when the residents were clearing their city of the bombing damage. The largest such Trummerberge is about 250 acres in size, and the back side of it is still being used for some commercial and industrial refuse, plus construction debris.

The study team visited Berlin's major landfill site, located in the southwest corner of the city, adjacent to the iron curtain separating the city from East Germany. This site receives all kinds of solid wastes--domestic, commercial, industrial, and construction. Approximately 25% of the total volume of West Berlin's solid waste is being buried there. The original site was an abandoned gravel quarry, but it appeared that the landfilling operation had overrun the old quarry. The landfill is surrounded on the West Berlin side by a forested greenbelt. The site has been used for 10 years, and the authorities figure it may suffice for 10 years more without too badly encroaching on the forested recreational area around it.

Except for the height of the refuse above the normal land elevation (about 30 feet), the landfilling operation appeared typical of many American operations. It could not be called a sanitary landfill because the refuse was not covered every day, but it was not an open burning dump, either.

Also in Germany, it is interesting that in 1969-1970 the large chemical manufacturer BASF at Ludwigshafen changed from incineration at the plant site for the disposal of all solid wastes, paper, wood, plastics, to controlled landfill. This has been done because of the urgent need to use incineration facilities for the disposal of other type wastes, chiefly liquids and pasty solids, to improve the situation concerning water pollution. BASF owns a large island some 15 miles upstream from the plant, and each day all solid wastes, including rubble, large solid objects and other building material as well as combustible items mentioned above, are placed on barges and transported to the island where the material is spread and compacted by steam rollers. Sufficient land is available for about 50 years of such controlled disposal.

In Sweden, a certain amount of open dumping takes place, but the largest amount of solid waste disposal (about 80% of the total) is by sanitary landfill. The Swedish authorities are currently making a systematic investigation of all of their counties to determine the best methods for collection, transportation and processing of solid waste from economic and aesthetic points of view. The general trend is toward more incineration and composting and away from landfilling.

In general, then, the situation in other countries as regards plastics in sanitary landfill is very little different from that in the United States. There is more actual effort abroad in the area of volume reduction prior to landfilling and some of the machinery developed there for pulverizing, grinding or size reduction is very advanced. Landfill principles for plastics disposal appear otherwise to be the same.

REFERENCES

1. Warner, A. J., C. H. Parker, and B. Baum. "Plastics Solid Waste Disposal by Incineration or Landfill," a study prepared for Manufacturing Chemists Association (December, 1971).
2. Staudinger, J. J. P. "Disposal of Plastics Waste and Litter," S.C.I. Monograph Number 35, published in England by Staples Printers, Ltd.
3. Jensen, M. E. "Observations of Continental European Solid Waste Management Practices," Public Health Service Publication No. 1880, Office of Solid Waste Management Programs, Environmental Protection Agency (1969).

4. Hart, S. A. "Solid Waste Management in Germany," report
    of the U.S. Solid Wastes Study Team Visit, June-July,
    1967, Public Health Service Publication No. 1812, Office
    of Solid Waste Management Programs, Environmental Pro-
    tection Agency (1968).

INDEX

INDEX

SOLID WASTE DISPOSAL

TABLE OF CONTENTS

VOLUME 2   REUSE/RECYCLE AND PYROLYSIS

Baum & Parker
March 1974